Scaffolding

SITE PRACTICE SERIES

General editors: Harold Lansdell, FCIOB, FCIArb,
and Win Lansdell, BA

Making and placing concrete – *Edwin Martin Baker*
Timber-frame housing – *Jim Burchell*
Scaffolding – *Richard Doughty*
Construction site security – *Len Earnshaw*
Industrial relations on site – *Tom Gallagher*
Site safety – *Jim Laney*
Fixings, fasteners and adhesives – *Paul Marsh*
Site engineering – *Roy Murphy*
Glazing – *Stanley Thompson*
Steel reinforcement – *Tony Trevorrow*

Scaffolding

RICHARD DOUGHTY

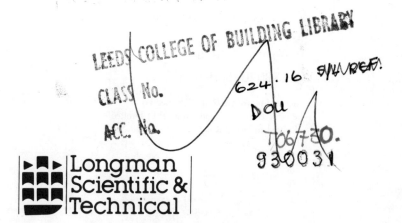
Longman
Scientific &
Technical

Longman Scientific & Technical
Longman Group UK Limited
Longman House, Burnt Mill, Harlow
Essex CM20 2JE, England
and Associated Companies throughout the world

First published 1986
Second impression 1988
Third impression 1991

British Library Cataloguing in Publication Data
Doughty, Richard
    Scaffolding. − (Site practice series)
    1. Scaffolding
    I. Title   II. Series
    690'.028   TH5281

ISBN 0-582-40625-0

Set in Linotron 202 10/12pt Bembo
Produced by Longman Group (FE) Limited
Printed in Hong Kong

Contents

Acknowledgements

We are grateful to the following for permission to reproduce copyright material; S.G.B. Group Plc., W. C. Youngman Ltd., British Standards Institution, 2, Park Street, London, W1A 2BS, from whom complete copies of the document can be obtained.

Author's Acknowledgements
My thanks also to Harold and Win Lansdell, and to Stewart Champion for his encouragement and support.

Introduction

'The men entrusted with the erection of scaffolds are seldom anything more than ordinary labourers, and as a rule wholly ignorant of the principles of Mechanics as taught in the School.' So said Hurst in 1871. He clearly would not have expected his ignorant labourers to be able to read correctly and interpret scaffolding drawings and specifications and then be able to erect the complex and designed scaffolds we are all familiar with today. This book is designed to cover the knowledge and practical understanding necessary for scaffolders, craft apprentices, lecturers, trades foremen, general foremen, safety officers, graduate engineers and anyone else whose work requires him to design, erect or inspect scaffolds.

Drawings, diagrams and easy checklists have been used extensively to avoid unnecessary verbal description, and information is presented concisely in a form readily accessible for both reference and practical purposes.

1

The history of scaffolding

There are many books on the history of architecture in which every method of building work or construction over the ages has been written about. A wealth of information is available regarding structures, materials, tools and methods used – everything except the means of support and access for the actual building process. Historians have ignored this particular operation, or have regarded it as unimportant except in the case of Stonehenge or the Pyramids.

To trace the development of scaffolding as we know it today, we begin in about the year 3000 BC in Egypt when quite ambitious structures were being built. The Egyptians, believing in life after death, thought that the body should be preserved in a lasting tomb. Their tombs therefore are among the first great constructions that were both solid and permanent.

There are 'graffiti' in rock carvings in the tombs of Amenehat and Senustret, two Pharoahs of the middle · Kingdom 2100–1700 BC, showing workmen standing on timber trestle scaffolds complete with ladders and handrail. Although it is difficult to be certain, it would seem that the lift heights were approximately 2 metres and the standard spacing 2.5 metres.

The superstructures of the royal tombs were brick with brick arches and wooden lined roofs. These were propped with poles during construction – in the same way as the Spanish still do – and illustrations of propping exist in the royal tombs at Memphis and Cheops.

Simple scaffolding was used on the internal tomb work, and rigs for beams for sarcophagus lids were erected using rope and pulleys. To give some idea of the tackle required, the granite beam in the King's Chamber of Cheops weighs 50 tonnes and was lowered 40 metres.

Construction methods evolved slowly through the Assyrian Empire, with its palaces built for its Warrior Kings, through the

Babylonian Empire, the Persian Empire and the Aegean civilization to the ancient Greeks, where timber scaffolds were used on temple work, and it is interesting to note that Hyclades, in 650 BC, stated that no temple should be built of greater height than surrounding cedar trees, as these were ultimately used as props to the lintels on the columns. The method was to erect a cedar tower interlaced with cross-members at approximately 5-metre heights, the tower being approximately 3 metres square. From successive lifts, a tackle was rigged and the circular column sections were hoisted until the tops were reached. After this, the spaces between columns were bridged and the lintels raised into position.

The next real step forward was under the great Roman Empire where systems of scaffolding evolved, particularly in the military field. All work outside Rome was done by the legions of the Roman army which employed indigenous artisans to erect all types of structures such as road bridges, forts, etc., and with typical military precision they perfected a standard method for all types of work. From wall paintings and mosaics, as well as written material, we know that their falseworks followed set patterns. The independent scaffolding towers constructed were very similar to those used up to this day and often a rapid conversion was done to turn them into embascules, war engines or entrilladas when attacked.

It is not until the time of the Norman Conquest and the rapid building of abbeys, priories, cathedrals and castles that we in England have any definite records of building work. Architects would vie with each other to build great monuments to God and new building methods were developed. Between the twelfth and fourteenth centuries the classical independent, putlog, birdcage and suspended scaffolds were evolved using timber poles and hemp lashings, and the terminology was evolved with the work it described.

The origins of many words are uncertain, but several have been used in their modern sense for centuries. For example, the word 'scaffold' itself: 'scaffotes' is found as early as 1349 in medieval Latin with its modern meaning and the Eton College accounts for 1442 include 'v. dosyn of hyrdelez for skafold'. We can read in 1471 of 'nedlis (needles) for scafelys', in 1466 of 'a purloyn on the sparres with punchions fro the bemes to bear the same'. 'Poinchon' is, in fact, already found with its modern sense in thirteenth-century' Old French.

By the early eighteenth century we find terms being precisely defined and so stabilized. So, for example, in 1703 'In Building of Scaffolds the Ledgers are those pieces that lie Parallel to the side of the Building'. The early manuals of practice at this time also contributed to the fixing of terms. There is still evidence of scaffolding to be seen on nearly every medieval building. Fifteen-centimetre stone inserts at approximately 1.5 metre heights on the walls were the putlog holes filled in. Above the triforium in Norwich Cathedral there are still visible the iron hooks from which a suspended scaffold was hung. There is a story concerning Norwich that in 1362 some work was being carried out to the main roof painting from suspended scaffold. Next to the cathedral was the Abbot's residence with a small hospital which got rather full of workmen who fell off the scaffold when stepping back to admire their work – a fall of some 30 metres. So serious was the problem that Abbot Jarviese hired from the parish for one angel (20p) per month an orphan boy named John Carrier aged nine, who was provided with a handbell which he rung every time a workman got too near the edge (the first recorded safety officer?). One medieval architect, Andrew Wykam, is on record as saying at the completion of one of his churches that he was 'full of woe' at the possibility of the spire and tower collapsing when the framework was removed, as the construction of the scaffold was far superior to that of the masons' stonework.

The first scaffold drawings produced can still be seen in the British Museum. These were by Christopher Wren for St Paul's Cathedral. From the 1750s onwards, a great deal of attention was paid to the design of falsework by engineers such as Palmer, Rennie, Telford, Tritton and Rendel, especially on bridgeworks. The day of the long span was here and the centering was a vital part. Consequently no engineer entertained the idea of any project unless he personally specified or designed all the elements and we are fortunate to have in existence some of the finest drawings of falsework and scaffolding from this period, most of which were drawn and coloured by the engineer himself.

By this time, specialist erectors were being used who went from project to project. This was, of course, the beginning of the engineering era and railways, mines, bridges and roads were providing a great programme of works.

As the canal builders, called navigators, were moving into the railways and being renamed 'navvies', so the temporary carpenters specializing in falsework and scaffolding were establishing their

3

trade. The first recording of 'scaffolders' as such was used on the construction of the Isambard Kingdom Brunel Temple Mead Station at Bristol in 1849.

During the 1850s, two principal methods of building were employed, known simply as the North Country system and the South Country system.

The North Country system involved the use of 'overhand' work, where platforms were laid on the floor joists inside a building and, as the work progressed, trestles were used by workmen until the next floor level was reached. (The external finish had to be achieved from the inside by reaching over the work, that is to say, 'overhand'.) So, in general, scaffolding was not required on the outside of the building.

The South Country system usually involved scaffolding being erected on the outside of a building and scaffolds were termed 'bricklayers' or 'putlog' and 'independents'. They were constructed of timber poles, platform boards and wedges and were lashed with three-strand manila or wire bonds.

The scaffolder's only tool for hundreds of years was a 'shingling' hatchet with a hammer head (see Fig. 1.1). With this tool he shaped the ends of putlogs, drove in wedges and nails (see Fig. 1.2), and, by twisting the cord around the middle of the handle, tightened the lashings (see Fig. 1.3), using the hatchet as a lever.

Fig. 1.1 Shingling hatchet

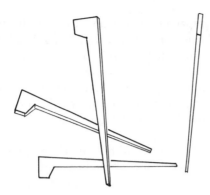

Fig. 1.2 Spikes

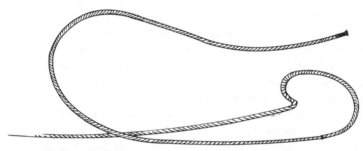

Fig. 1.3 Wire lashing

It seems strange that up until the 1860s there was no specialist company dealing with scaffolding, the builder's carpenter usually doing the bulk of the work. Several companies lay claim to being the first specialist scaffolding company, but it is recorded that Tasker and Booth and James Stephens both entered the field of scaffolding specialism in about 1863. In 1880 Edwin Palmer formed a company mainly for cradles and suspended work, starting conventional scaffold work around 1890.

It could be said that the period between 1870 and 1910 was the heyday of timber scaffolding. By the turn of the century the iron and steel age was established, and although tubular steel scaffolding was first used in 1896 it took something like fourteen years of experimentation to arrive at the 48 mm diameter tube. A whole new range of products was brought out to match. Scaffold couplers were first patented in 1911 with hop-ups, towers, catheads and props, and by 1920 a fair proportion of scaffolds was being erected in steel tube. New companies were being formed. In 1904 a company called 'OSSA' was using tubular scaffolding. Big Ben Ltd and London Scaffold experimented with tubular scaffolding. S.G.B. (Scaffolding Great Britain) was formed around 1910. The use of timber was rapidly diminishing and a disastrous fire at Lockwood's premises in the Strand in 1927, caused by the timber scaffold catching fire, prompted the greater use of tubular scaffolds, certainly in cities.

In the USA a great deal of work had been done on tubular steel scaffolding since 1910 by the Patent Scaffold Company and the American Bravo Company. Even so, up to 1929 the majority of scaffolds in the United States were timber, even for the new skyscraper buildings.

In 1927 the Sherry Netherlands Hotel was severely damaged during construction in New York when the timber scaffolding around the 38th storey burnt out over a period of twelve hours. This was virtually the end of timber scaffolding in the cities of the USA, and it did not help other forms of scaffolding either, as from then onwards most of the development swung towards more sophisticated suspended equipment and powered winches.

By the outbreak of the Second World War, most of the large firms were established and most scaffolding work was being carried out in tubular materials; but the standards and methods of construction were still those laid down in the days of timber.

A series of disasters and collapses occurred in the ensuing years. Too much reliance was being placed on experience and a 'good eye' alone. Failures of scaffolds were becoming more frequent as the work undertaken became more and more complicated. The whole concept of scaffolding had changed from the traditional jobbing access structure to an engineering concept, but no design facilities or planning existed and there was no recorded standard practice.

The outbreak of war in 1939 provided a respite for scaffolding companies. Most of them were engaged on shelter shoring and a large proportion of materials was used on tank traps, beach defences and even taken to France on D-Day to help the temporary harbour works and bridging jobs.

Then came the end of the war and few building materials, steel, wood, concrete, etc., were available. Yet a massive rebuilding programme was needed. The scaffold companies had their materials back and found themselves in the position of having to supply and erect all manner of structures such as bridges, footbridges, water towers, grandstands, and types of structure which would normally require detailed design and engineering experience.

It was during the period 1945–50 that the majority of large companies set up engineering departments and employed design staff. A proper engineering approach began to be taken but sometimes the old traditional methods and outlooks were still applied in structures which should have been designed and this sometimes led to collapse.

In addition to this, the work to be carried out by scaffolding companies was allotted neither enough time nor money in the preparation stages of the main project. On an engineering permanent works project, a consultant would be allowed anything up to two years for the preparation of designs and drawings, while the scaffolding companies were expected to produce their designs and proposals in some cases in a few days.

In 1943 a group of scaffolding companies formed the National Association of Scaffolding Contractors (NASC). Membership was offered to companies and firms which hire and/or erect and dismantle scaffolding in the United Kingdom as a major activity of their business. It elected as its first president Mr Clifford Jones. While concerned with the promotion of scaffolding contractors' trade and other interests, the NASC also aims for the maintenance of a high standard of quality, design and workmanship in the manufacturing and maintenance of scaffolding equipment. It promotes and finances research, and its council has various specialist committees and working groups. The NASC took an active part in introducing the training certificate scheme for scaffolders.

The scaffolding industry has made a significant contribution in the field of engineering over the past hundred years, and some of the work undertaken is worthy of recording in line with permanent structures. It is now recognized as a specialist service and duly respected by the major professional institutions as such.

In the last decade the scaffolding industry has been losing its traditional image and gaining the professional status of consultants in temporary works.

Running in parallel with this development, the statutory instruments of Government have been codifying the requirements of working platforms. These, together with Codes of Practice, have helped to improve the safety of scaffolding.

2

Safety

General principles

Every year there are many men killed or injured in accidents connected with scaffolding. It is estimated that there were, between 1978 and 1983, something like 3000 accidents involving falls from heights, including 30 fatalities.

The three principle requirements for scaffolds are that they must:

(a) be suitable for their purpose;
(b) be safe;
(c) comply with the regulations.

Suitability

To be suitable for its purpose the scaffold must provide adequate space:

(a) for workmen to come and go;
(b) for the movement and stacking of materials; and
(c) for the work to be done conveniently.

It must be strong enough to carry the weight of the men, materials and plant used.

Safety

This is the overriding consideration. For the scaffolder it is not sufficient to provide a safe structure for others to work from. He must look after his own safety whilst working, that of those working with him, and that of those nearby, either on the site or close to it, who might be accidentally hurt.

For scaffolds which are erected from the ground and which progress upwards, safe working procedures are relatively straightforward, but for those which suspend either from the sides of buildings or beneath ceilings or soffits, the scaffolder must not attempt to emulate a trapeze artist or be expected to sprout wings like a fairy.

If there is no other way, the scaffolder must construct his own safe working platform up from the ground or from a barge under a river bridge to get up to make and fix to slinging points. Only very experienced scaffolders should attempt the tricky and precarious steps sometimes necessary, and the appropriate precautions using safety nets or harness or both should always be taken.

Regulations

Those which govern scaffolding are the Construction (Working Places) Regulations 1966 containing thirty-nine detailed regulations. These are not easily interpreted into everyday working language and have been expressed more simply in a publication *Construction Safety* by the Building Advisory Service and reproduced in Chapter 14 of this book. This publication was an attempt by the BAS and others, including the CITB, to help those with site responsibilities. The booklet can be obtained from the BEC or the BAS.

At the CITB Training Centre at Bircham Newton, Norfolk, the common types of scaffold and the procedures to erect them have been taken and developed to evolve quicker, safer and easier methods of satisfying the three principal requirements. These methods are incorporated in this book. If they are adopted and the regulations complied with, the finished scaffolds will have been safely erected and will be safe to work on. Further advice on safety on the building site generally and with particular reference to scaffolding can be found in the Site Practice Series book *Site Safety* by J.C. Laney.

Safety elements

Elements common to all scaffolds and which contribute to their safety are: soleplates; ledgers; transoms; bracing; ties.

Soleplates

Soleplates are of timber or other suitable material, of sufficient size to distribute loads from the baseplate to the ground or other load-bearing surface.

All soleplates must be bedded throughout their full length. They must be sound and preferably so placed that each piece takes the load from two or more standards. Any junction or joint in a soleplate must be set within the middle third of the distance between two standards, and the soleplate must continue to at least the same distance beyond the end standards. Some of the areas where it is advisable to have soleplates placed are:

- slate roofs, tiled roofs, asphalt roofs, bitumen felt roofs;
- ordinary ground, grass, asphalt paths, sloping ground, made-up ground, etc.;
- mosaic, parquet and other finished floors.

Ledgers

Ledgers are longitudinal tubes usually fixed parallel to the face of the building in the direction of the length of the scaffold. They also give support to other parts of the scaffold, such as transoms, putlogs, puncheons, hanging tubes, etc.

They should be level and fixed to standards with right-angle couplers.

Joints in ledgers adjacent to each other should be staggered. Joints in ledgers should be secured with sleeve couplers, which should fall to within a quarter to one-third of the distance between two standards – the ledgers should never be joined in the middle of the bay.

The distances between lifts are specified in BS 5973.

Transoms

Transoms are tubes which are fixed horizontally across the ledgers between the outside and the inside ledgers. They are used to secure the standards and may act as board supports.

The spacing of transoms will depend on the load they are to carry and the thickness of the boards.

Transoms may be fixed to standards with right-angle couplers or to ledgers with putlog couplers. If fixed with putlog couplers, they should be fixed as close to the standards as possible and no more than 300 mm away. The main transoms must never be removed without authority. Intermediate transoms may be removed temporarily if necessary, providing they are replaced when the scaffold is being dismantled. Care must also be taken that transoms do not project beyond the ledger, especially in the first lift where they could constitute a hazard. It is sometimes necessary to connect longitudinal or sway bracing to transoms, and right-angle couplers must be used for this.

Statutory regulations for transom spacing when supporting boards are:

Board thickness	Maximum transom spacing
38 mm	1.5 m
50 mm	2.6 m
63 mm	3.25 m

Bracing

A scaffold consists of a number of squares and without bracing any movement in the scaffold tends to throw the scaffold out of square.

In order to stiffen the scaffold and resist the possibility of it becoming out of square, the squares must be converted into a series of triangles by fixing braces. Ideally the braces should be fixed to the ledger, as close as possible to the standard and ledger connection, with right-angle couplers. As close as possible, in this case, means within 300 mm of the connection. The spacing of braces is important, and braces should be fixed at alternate pairs of standards, but it is equally important that the ends of the scaffold are never left 'open', i.e. both ends of the scaffold must always be braced and every alternate pair of standards in between, even if this means two braces fixed adjacent to each other.

There are three generally accepted ways of bracing (see Fig. 2.1) but the direction of the bracing is not important. It is far more

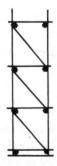  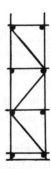

Fig. 2.1 Bracing

important to make sure that the bracing is in position, rather than demand that a particular method of bracing is used.

Ledger bracing should be fixed from ledger to ledger with right-angle couplers, except where a lift has to be boarded out. In that case the top of the brace should be fixed to the ledger with a right-angle coupler, but the bottom of the brace fixed to the standard with a swivel coupler, just high enough on the standard to allow the boards to sit properly between the standards.

Where it is necessary to erect scaffolds on pavements where bracing is restricted or not at all possible, ledger bracing may be omitted, but a word of warning here: the unsupported length of the standards must not exceed 2.6 m. If this figure is exceeded a drastic reduction in the loading capacity will take place. If the lower lift does exceed 2.6 m, knee braces should be fixed, starting just above head height, say 1.8 m, to the top corner of the lowest lift. The knee braces must be fixed on every pair of standards. It is also advisable to make sure that ties are started at the lower lift at every third pair of standards at least (see Fig. 2.2).

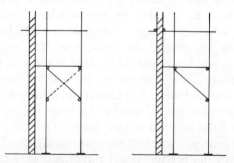

Fig. 2.2 Pavement bracing

Sway bracing or longitudinal bracing may take the form of a zigzag from the base to the full height of the scaffold, or a continuous slope from bottom to top (Fig. 2.3). If this latter method is used, the brace must be fixed at an angle of between 40° and 50°, fixed to the standards with swivel couplers, as near to the standard to ledger connections as possible, or fixed to extended transoms with right-angle couplers (Fig. 2.4).

If bracing is to be fixed to extended transoms with right-angle couplers, then the transoms themselves must be fixed to the standards with right-angle couplers. It would be silly to fix bracing to transoms which in turn were only secured with putlog couplers.

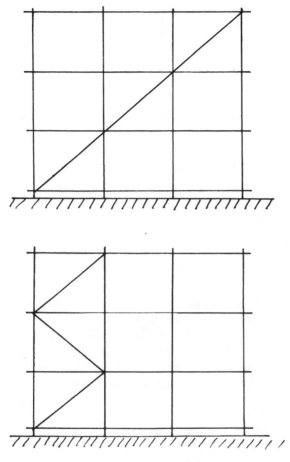

Fig. 2.3 Sway bracing

Sway bracing must be provided to all scaffolds where there is movement along the face of the building and the scaffold is not prevented from moving by any other means.

Where a scaffold is erected between buildings which afford opposing surfaces to butt the scaffold against, it may be acceptable to leave the sway brace off providing the space is not greater than 10 m, but again, the scaffold must be tied at least at every alternate lift.

If there is still movement in the scaffold, plane bracing (i.e. a brace in a horizontal plane) should be inserted, fixed to two-way or through ties.

13

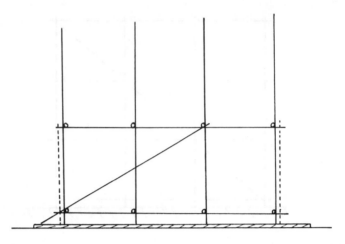

Fig. 2.4 Sway brace

Ties

Ties are for tying scaffolding into building facades and should be
provided to resist inward and outward movement of the scaffold.
Where possible, they should be fixed in such a way as to provide
restraint additional to that obtained from the sway bracing. The
strength of the building structure at the tying point must be
sufficient to take any loads that may be transferred to it.

Box ties

These consist of a number of tubes and right-angle couplers
arranged to form a square fixed around columns of the building to
resist both inward and outward movement and provide some
lateral restraint (Fig. 2.5). They should preferably be placed just

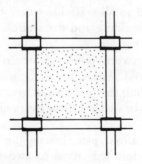

Fig. 2.5 Box tie

14

below the scaffold lift and fixed to the standards with right-angle couplers.

Through (or two-way) ties
These are more commonly known as two-way ties, and their strength lies in their push–pull effect. It is important to remember that scaffolds can deflect inwards, as well as outwards, and scaffolders should always butt transoms hard against the wall on every lift.

Through ties rely on tubes fixed across an opening (such as a window or door) on both the inside and the outside. It is a good idea to fix the inside tube *vertically* resting on the floor so that it cannot slip downwards, but it may be fixed horizontally if this is more convenient (Fig. 2.6).

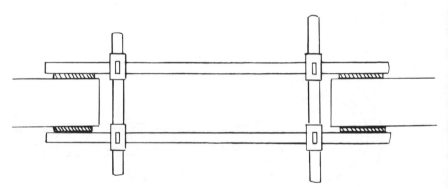

Fig. 2.6 Through (or two-way) tie

If this method is preferred then it is important to ensure that the tie tube (or tubes) attached to the horizontal tube is as close to one of the edges of the opening as possible (Fig. 2.7).

Bolt ties
These ties are obtained by drilling a hole into the structure to which the scaffolding has to be tied. An expanding anchor-bolt is fixed which grips on expansion (Fig. 2.8).

Some bolt ties are threaded to receive a ring bolt through which a scaffold tube may be passed, or to which a tube may be tied. Other bolt ties have ordinary nuts, bolts or studs for fixing a short, specially made, tie tube.

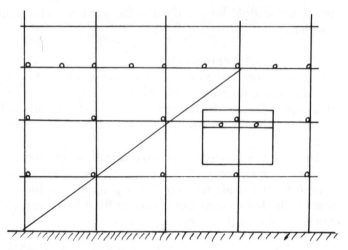

Fig. 2.7 Position of ties in window opening

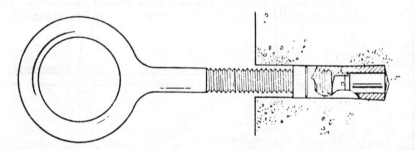

Fig. 2.8 Bolt tie

There are several proprietary makes of bolt tie, but the main advantages they share are:

    (a) as many tie points as needed can be provided, irrespective of available openings;

    (b) tie points can be placed as required within the limits imposed by the structure;

    (c) fewer ties may be required as they do not have to be temporarily removed;

    (d) there is a high safety factor;

    (e) bolt tie anchor points can be used over and over again.

Reveal ties

A reveal tie is a very useful tie and one which, if used properly, can quickly be fixed thus saving time and money.

A reveal tie is a tube which is wedged into an opening in the wall by means of an expanding type pin known as a 'reveal pin'. A tie tube is then fixed to the tube wedged across the opening (Figs 2.9 and 2.10).

Fig. 2.9 Reveal tie

Fig. 2.10 Reveal pin

Reveal ties must be inspected every seven days because vibration may have the effect of loosening the reveal pin, or the timber packing may have shrunk due to sunshine or drying winds. It is common for scaffolders to use off-cuts of scaffold boards for packing but this is inadvisable because scaffold boards are soft wood and a good deal of swelling and shrinking takes place.

Ideally, hardwood should be used as packing, about 5–10 mm thick. If this is not available on site, then the thinnest packing available should be used and either plywood or hardboard is quite acceptable. The packing is used for protection of the building fabric forming the opening, and to secure a better grip than that obtained by metal on to a masonry surface.

It is not good practice to use reveal ties on putlog scaffolds as bricklayers will not be pleased if pressure is put on their green brickwork. Through ties should be used instead and they should be well behind the mature brickwork.

Hook ties

In certain circumstances it may be possible to form an L-shaped hook with tube and fittings to hook over or around part of the building, but care should be taken if using this method, particularly if the tube is hooked over a parapet. Parapets which have weathered are not always the firmest features to tie to.

3

Equipment

Tubes

Inspection

As the stability of a scaffold is largely dependent on the strength and condition of the tubes used in its construction, it is most important that tubes are inspected before they are used.

The main points which should be checked are:

(a) Tubes must be straight throughout their length.
(b) Tube ends should be cut clean and at right angles to the axis of the tube. If tubes are not cut square they are liable to split when used as standards.
(c) Tubes should be free of rust corrosion, flaws, splits, dents and other visible damage.
(d) Tubes should have a suitable protective coating. Unprotected tubes should not be used in water and particularly not in marine structures.

There are generally two types of tube used in the United Kingdom: steel and aluminium alloy. In the main, of course, mild steel is used and the tube specifications regarding dimensions and quality are to be found in BS 1139 (metal scaffolding). Aluminium alloy tube specifications are also contained in BS 1139.

Tubes are normally supplied in 6.3 m lengths. Shorter lengths are supplied to order.

Storage

Wherever possible tubes should be stacked in respective lengths and stored in racks. If racks cannot be provided then tubes should be formed into pyramids in sizes. Both methods make identification and selection easier, providing tubes are stored with their ends flush.

Good housekeeping is an essential part of the care of all plant and the following points are worth noting.

1. Before corroded tubes are stored they should be wire-brushed, the extent of the corrosion assessed by an expert examiner, and if found with excessive corrosion or deterioration they should be scrapped.

   The examination of corroded tubes to determine the possible loss of strength should be left to the expert. However, a rough and ready guide that can be used is to cut 300 mm of a suspect tube and weigh it. It should not weigh less than 90 per cent of its minimum specified weight.

   Although tubes are normally hot-dipped galvanized, there are still plenty of steel tubes around to justify a good wire-brushing.

2. Racks for storing tubes are usually designed in the form of a birdcage with bay lengths and lift heights suitable for the sizes of tubes to be stored. Care should be taken when loading any bay not to exceed the safe working load of any one coupler. Although safety check couplers are often seen supporting the coupler, it must not be assumed that each additional coupler will have the same working load.

   To allow for the impact loading imposed on the rack, i.e. the inevitable banging and crashing of storage and removal, add 25 per cent when calculating the maximum individual load to be deposited.

   The design of the rack should take into account any wind forces on the racking itself if provided with sheeting cover.

Corrosion

Steel tubes have an average life of between ten and twenty years, but this can only be a very rough guide. There are many factors affecting this, particularly atmospheric conditions and the degree of exposure in those conditions. For instance, it is quite possible for tubes to be standing for one or two years in a sulphurous and salty atmosphere or extremely wet conditions. It will be appreciated that the corrosion rate may be such that it could reduce the structural safety considerably. It is quite possible, for instance, to find that tube corrosion does not form uniformly over the surface of a tube and it may be that corrosion has also taken place inside the tube.

The couplers may contribute to local corrosion if small pockets of moisture collect between the coupler and the tube. Local corrosion can also take place when two dissimilar metals are in contact with each other (e.g. alloy couplers used with steel tube) a white powder may be seen around the coupler.

To assess whether a corroding tube has become unsafe to use, a rough guide is that a tube with only half its original wall thickness is only half as strong as when new.

Couplers and other fittings

The idea of tubular steel scaffolds was first put forward in 1896 when the famous engineer Isambard Kingdom Brunel designed a system of scaffolding for the building of the 'Great Eastern' steamship. The drawings and the patents are now in the Institute of Civil Engineers' library. The scaffold was never built, other than a hand-forged prototype, because of the work entailed in forming the joints and the inconsistency of the available materials. There followed a great deal of experimentation over the next fourteen years. Combinations of timber poles and steel 4 in. cast-iron tubes and all kinds of weird and wonderful iron couplers were patented, and among those couplers was a fitting invented in 1896 called the chain and bracket (Fig. 3.1).

Fig. 3.1 Chain and bracket – an early, iron scaffold coupler

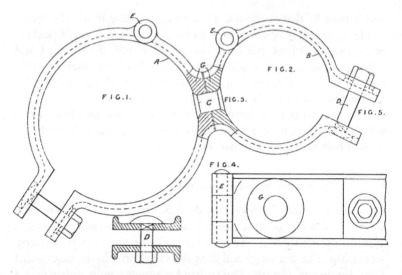

Fig. 3.2 Probably the first one-piece scaffold coupler used to couple two larch poles either as standards or as ledgers by gripping the thick end of one pole and the thin end of another

The first one-piece coupler was probably the device patented in 1903 by Edward Wilding, a blacksmith (Fig. 3.2), who claimed that it was a 'quicker and a more safer way of erecting scaffolding poles together by means of a pair of wrought iron steel or malleable iron clips for use in the erection of all kinds of scaffolding poles'.

Couplers

A coupler is a device for coupling together two tubes in any position.

There are four main types of coupler in use in the general construction of tubular scaffolds:

    (a) right-angle couplers;
    (b) swivel couplers;
    (c) joints (sleeves and spigots);
    (d) putlog couplers.

Each of them is to be found in a great variety of designs according to their makers. The majority of couplers used in the United Kingdom are made from steel or malleable iron, although there are some in use that are made of aluminium alloy.

Couplers are generally classified as:

(a) spring-steel;
(b) drop-forged mild steel;
(c) pressed-steel.

Spring-steel couplers have their components fixed rigidly together and these can be tightened by the nut and bolt action bending them. Drop-forged and pressed-steel couplers require a hinge to allow the components to be pulled tight.

Right-angle couplers

Spring-steel right-angle coupler. This particular coupler (Fig. 3.3) is designed to be used 'coupler on coupler', i.e. two right-angle couplers are used to secure a ledger and a transom to a standard, and in doing so allow the transom to sit directly on the ledger, thus positioning the transom on the same level as the intermediate transoms. This is very useful when it is necessary to have all the transoms at the same level in order to carry the working platform boards.

Another feature of this right-angle coupler is that it allows the construction of very strong node points; using 'coupler on coupler' gives twice the safe working load.

This coupler may also be used as a three-way fixing where great rigidity is required by fixing, in addition to the two couplers securing the ledger and the transom to the standard, a third coupler securing the transom to the ledger.

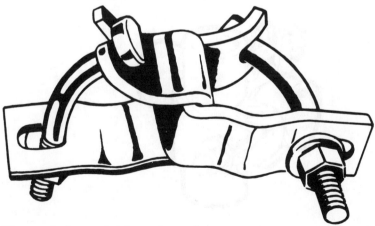

Fig. 3.3 Spring-steel right-angle coupler

Drop-forged right-angle coupler. There is a wide range of shapes and strengths in the drop-forged category (Fig. 3.4).

The advantage of using drop-forged couplers is that they are very strong, very rigid and generally very reliable. Some of the drop-forged couplers have a safe working load far in excess of the BS requirements.

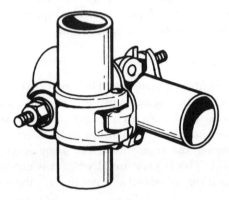

Fig. 3.4 Drop-forged right-angle coupler

Pressed-steel right-angle coupler. Pressed-steel couplers (Fig. 3.5) are fairly recent additions to scaffolding. They are quite cheap and simple to manufacture, but in every respect meet the minimum requirements of BS 1139.

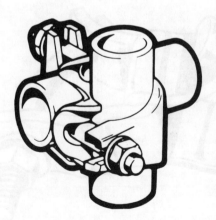

Fig. 3.5 Pressed-steel right-angle coupler

The universal coupler

The universal coupler (Fig. 3.6) is more commonly known as a band and plate. The advantages are: firstly, its versatility – it can be used as a right-angle coupler, a parallel coupler, or a putlog coupler; secondly, it is generally quick to secure, having only one screw plate to tighten; and thirdly it may be easily maintained.

However, the disadvantages are that it is bulkier than most couplers, over-tightening the screw plate will dent the tube, and it has loose parts, i.e. the 'band', the 'plate' and the 'chair'.

WARNING! It is not good practice to mix different makes of coupler on one job, as some couplers are incompatible with other makes. Scaffolders then have to carry more tools, which may lead to lower productivity or workmanship. It is better to use one make of coupler rather than buy a job lot of mixed scaffolding couplers.

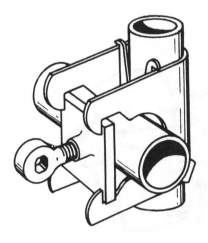

Fig. 3.6 The universal or band and plate coupler

Other fittings

Swivel couplers

Swivel-up couplers are used for joining tubes at an angle other than a right angle (Figs 3.7 and 3.8).

Joints

(a) *Spigots*. Also known as joint pins, these are internal fittings to join one tube to another coaxially (Fig. 3.9).

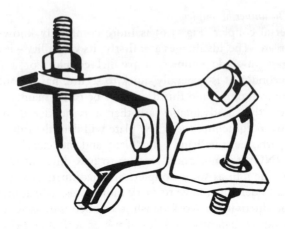

Fig. 3.7 Spring-steel swivel coupler

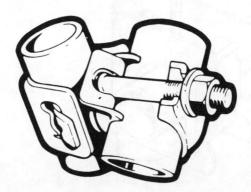

Fig. 3.8 Pressed-steel swivel coupler

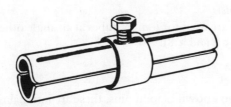

Fig. 3.9 Spigot joint

Fig. 3.10 Sleeve coupler

(b) *Sleeve couplers.* These are external couplers used to join one tube to another coaxially (Fig. 3.10).

When used other than vertically, spigots and sleeve couplers should be arranged so that they are never more than a quarter span away from any standard. They should never be placed at mid-span.

Putlog couplers
Putlog couplers are used for fixing putlogs or transoms to ledgers (Fig. 3.11).

Fig. 3.11 Putlog coupler

Baseplates
Although strictly speaking a baseplate is not a coupler, it is an important fitting and should be inspected and used correctly.
   There are three main functions of a baseplate (see Fig. 3.12):

(a) Distributing the load from a standard or raker;

Fig. 3.12 Baseplate

    (b) preventing lateral movement;
    (c) preventing damage to the tube.

Baseplates must measure at least 150 mm × 150 mm and, if the plate is mild steel, it must be at least 5 mm thick. If it is made of any other metal, then it must be thick enough to be capable of distributing the load from the standard. If fixing holes are incorporated into the plate, they must be positioned opposite each other, 6 mm in diameter, not less than 50 mm from the centre of the plate, and not less than 19 mm from the edge.

The shank must be at least 50 mm long.

In many situations it will be necessary to support the baseplate on a soleplate.

Soleplates

This is timber or other suitable material, of an adequate size and capable of distributing loads from the baseplate to the ground or other load-bearing surface. Note the use of soleplates as described in Chapter 2.

Inspection of couplers

Defective fittings must not be used, so couplers must be inspected before use.

Couplers used for scaffolds must be in good condition and free from corrosion and other defects which could affect safety. They must be inspected to ensure that there are no: (a) stripped threads; (b) distortion; (c) seized components; (d) rust.

Checklist for different couplers

Type	Use	Points to check
Right-angle coupler	To connect two tubes at right angles.	Right way up. Square on tube. Tightened correctly.
Swivel coupler	To connect two tubes at any angle other than a right angle.	Square on tube. Tightened correctly.
Putlog coupler	To fix a putlog or a transom to a ledger.	Square on tube. Tightened correctly.
Sleeve coupler	To connect two tubes when extended vertically or horizontally, an external coupler.	Tubes butted on centre plate. Tightened correctly.
Joint pin/spigot	To connect two tubes end to end internally.	Seated correctly. Tightened correctly.
Baseplate	To distribute the load from the standard.	Adequately supported on soleplate.
Toe-board clip	To attach toe-boards to standards and prevent accidental displacement.	Right way up. Correctly seated.
Putlog adaptor	To fix to the end of a tube to form a putlog.	Right way up. Positioned correctly on tube.
Reveal pin	To secure a tube between opposing surfaces.	Seated correctly in tube. Suitable timber packing. Tightened correctly. Checked at intervals for tightness.

Boards

Scaffold boards, scaffold planks, scaffold battens, are all acceptable terms.

For this chapter the term 'boards' will be used.

Boards should be made from soft wood, sawn finished on all surfaces and must be suitable for use with metal scaffolding. They should be capable of withstanding a load of 6.7 kN/m^2 when supported at 1.2 m centres. The ends should be bound with hoop iron 25 mm wide and at least 0.9 mm thick extending to at least 150 mm along each edge. The overhang of a board beyond its last support must not be less than 50 mm nor more than 4 times the board's thickness, e.g. when using 38 mm boards, the maximum overhang is 4 × 38 mm = 152 mm, so a useful tip is that if the transom is fixed at the ends of the hoop iron the regulations will be complied with.

To comply with BS 2482 the hoop irons should be either sheradized or galvanized. The hoop irons must be secured at each end of the board and on both its edges with large-headed clout nails at least 30 mm long, and must be finished off to avoid injuries.

The dimensions of the standard scaffold board are:

- Width: 225 mm plus or minus 6 mm
- Thickness: 38 mm plus or minus 3 mm
- Length: 3.9 m plus or minus 50 mm of the stated length.

The following timbers are generally used for making scaffold boards: Spruce (European or Eastern Canadian); Douglas Fir; Whitewood; Redwood; and, rarely, Pitch Pine.

Scaffold boards may be made from timbers other than those mentioned if the user is satisfied that they are suitable. In general, all timber used for scaffold boards must be free from any splits, checks, shakes, damage or decay which may affect the strength of the board.

The density of the board is also important. For instance, when handling a board, if it feels light, i.e. about 25 per cent lighter than the average board, it should be regarded as not of adequate strength.

Inspection

When inspecting boards for use, look at:

(a) The face: not more than one-third of the width of the

board in any one place should be knot-wood; the face of the board at the end should not be split more than 300 mm with the hoop iron fixed.

(b) The edge: the grain should not cross from face to face of the board in a distance of less than 300 mm.

(c) The length: along the board from one end, it must not be warped by more than 25 mm.

(d) Boards which are split, decayed or warped must not be used; but they need not be scrapped. The affected parts may be cut off to make shorter boards, and the newly cut ends of the boards must be protected with hoop iron.

Boards **must not be painted** as this may conceal defects.

Non-standard size boards may be used, e.g. boards 51 mm thick, which should be not less than 210 mm wide, and boards exceeding 51 mm thick, which should not be less than 155 mm wide.

Support

It is important that the spacing of the intermediate transoms (the board-supporting tubes) is correct. Working platforms should be close-boarded, each board having at least three supports, unless the span or thickness of the board is such as to prevent undue sagging. Such boards should be supported at centres not exceeding 1.5 m.

Decking

Boards should be butted and not lapped, i.e. boards should be placed flat with the ends of the boards butted forming a level working platform and not lapped one over the other, giving rise to a possible tripping hazard. If lapped, a bevelled piece of timber to reduce the tripping hazard should be fitted.

Handling

It makes sense to raise or lower scaffold boards by means of a gin wheel and rope or a light line, but on occasions it may be necessary to pass scaffold boards up or down a scaffold hand over hand.

If this is the case, the following precautions are necessary:

(a) A board being passed to someone else **must not be pushed** nor released until the person receiving the board has control of at least two-thirds of it, or says that he has it under control.

(b) The board must be gripped firmly and passed in a smooth movement.

Ladders

From the earliest time, ladders have played an important part in the history of mankind – symbolically, as in the dream of Jacob, when the Angels ascended to and descended from heaven by means of a runged appliance; and actually, when old-time warriors climbed the ramparts of beleaguered citadels by means of scaling ladders. Throughout the ages the ladder has been a symbol of progress. To get their feet on the lowest rung and climb the ladder of fortune, has been the aspiration of all who aim to succeed through their own exertions.

Ladders in themselves possess a peculiar significance, being regarded by not a few of even the present generation as harbingers of luck or otherwise, and many seek to avoid passing under a ladder at all costs, as the act of doing so will bring 'bad luck'. Perhaps there is some rational foundation for this superstition.

Ladder accidents

There is now a change taking place on construction sites with regard to safety. The Health and Safety at Work Act 1974 places the weight of law behind efforts to make the construction site safer. It has been estimated that between 1200 and 1400 ladder accidents happen in the construction industry alone each year.

Studies have been made and show that a large proportion of these accidents are caused by ladders slipping or not being adequately secured. Other typical accidents are: the user of the ladder slipping on rungs, or missing his footing, losing his grip, overreaching or overbalancing.

The following tables are taken from the Department of Employment and Productivity publication 'Accidents in the Construction Industry – Falls from Ladders'.

Analysis by occupation

The highest number of accidents occurred to painters (40%). Other activities involved were roofing (10%), guttering and pipework (9%), woodworking (9%), concrete and brickwork (9%).

Table 3.1   Analysis of cause of ladder accidents

Unsecured ladders	Overbalancing or overreaching	Slip on rung	Defective ladders	Total
204	55	84	40	383

Table 3.2   Use of ladder when accident occurred

	Unsecured ladders	Overbalancing or overreaching	Slip on rung	Defective ladders	Total
Ladder used as working place	110	35	16	14	175
Ladder used as means of access	89	17	68	22	196
Use not known	5	3	—	4	12

One of the reasons for so many accidents relating to ladders is that they are used to do jobs for which better equipment is available, e.g. trestles, stagings, towers, hydraulic lifts, etc. Ladders are really only suitable:

(a) to enable a person to rise or descend from one working level to another;
(b) for one person at a time;
(c) where the extra load due to materials or tools does not generally exceed the carrying capacity of one man;
(d) where use is likely to be short-term or intermittent only;
(e) over moderate distances between levels and at modest heights;
(f) where they can be deployed safely and made secure.

Types of ladder

There are many types of ladder in use today and they can be broadly categorized as follows:

33

Steps

The conventional step-ladder has rectangular stiles and flat, rectangular treads which are arranged to be horizontal when the ladder is being used. However, some types use flat-topped rungs.

Others have twin, tubular metal stiles and treads supported by tube or rectangular section.

Steps can be:

(a) *Lean-to* – designed to lean against a surface and not capable of self-support.

(b) *Travelling* – having hooks, slides, wheels or carriages enabling the steps to travel along tubes or rails.

(c) *Swing-back* – with a hinged, swing-back section to enable the steps to be self-supporting. Struts, cords or chains limit the movement of the swing-back section to give the correct spread at the base and deploy the steps at the correct angle.

   Where treads or rungs are incorporated in the swing back section, the ladder becomes a double-sided step ladder and one side may be capable of extension.

(d) *Platform* – incorporating a swing-back section and a platform to stand on. On very short steps and step-stools the platform may be at the top, but for steps of any length the platform should be below the level of the heads of the stiles so that the higher hand-hold is available.

Trestles

Trestles, as used by builders, are similar to swing-back steps but are designed so that a pair can bear the load of a light working platform not exceeding about 3 m in span. The stiles are a little narrower at the top than at the base and support cross-bearers, which are like the rectangular treads of a normal step-ladder turned through 90°.

The treads are normally spaced about 450 mm apart, twice the normal separation of ladder rungs. Trestles are essentially heavy-duty implements and are strongly built and reinforced with tie rods.

Standing ladders

This category includes most conventional single-stage ladders having rectangular stiles with rungs which are basically rectangular, circular or box-section.

Extension ladders

These comprise two or more standing ladders held one against another so that by relative sliding movements of the sections various heights can be reached. This type of ladder is useful where storage, access or space is too limited for a long single-section ladder.

In most types, the sections are graduated in size so that the stiles of one section fit inside those of another, thus providing resistance to sidesway when the ladder is extended. In the extended position, the smallest section becomes the highest. However, in some types the sections are of equal size and fit one against another so that the back section is raised. This reduces the risk of a man losing his footing at the overlap.

Extension is carried out in one of two ways:

(a) *Push-up type* – With the shorter types, mainly in two sections, extension is achieved by pushing up the top section, which is then held in position by fixed or swivelling latching hooks. These may be fixed with a safety lock. The extension is limited to a man's reach (say 2 m) unless the ladder is laid horizontally, set to the required extension and then reared.

(b) *Rope and pulley operated* – Longer multi-stage ladders are extended with the aid of rope and pulley mechanism.

Fixed ladders

Where it is not possible to provide a stairway, or high rise is required with safety, fixed metal ladders can be used.

These are generally of steel with rungs welded to the stringers. Suitable rest platforms must be fitted at least every 10 m. Hoops should be fitted at suitable intervals for safety.

Electricians' ladders

Electricians often use general purpose ladders, but prefer lightweight types which are safe to use in the vicinity of live electrical conductors. Metal and alloy are dangerous and are not recommended even where insulators are provided. Wooden ladders are dangerous when wet. Ladders of extruded glass-fibre provide a high degree of safety for electrical work.

Pole ladders

These ladders have half-round pole stiles (each stile being half of a long whitewood pole) and round rungs which are hand-made or turned.

The stiles These must be straight-grained and relatively light for their strength. Suitable woods include Spruce, Douglas Fir, European Whitewood and European Redwood.

The rungs Traditional rungs are wooden and round. In high-quality timber construction, hardwood rungs are individually tested, then shouldered, glued, force fitted and pinned into the. sides of the stiles without passing through them. Where, in some types, rungs extend right through the stiles, resistance to loosening can be provided by driving in hardwood wedges. A modern resin glue is normally used to provide durable security and prevent ingress of moisture into the joint.

Care of ladders

Because of the high capital outlay of ladders, e.g. a large construction site may have scores of ladders in use, it is important both economically and for safety that proper care of ladders is taken.

Storage

Ladders are often thrown carelessly to the ground, exposed to weather, rain and sometimes impact damage. Ladders should be supported horizontally clear of the ground on an adequate number of supports (ladders over 7 m in length should have at least 3 support points to avoid sagging). Do not hang a ladder by its stiles from spikes fixed to a wall.

Storage should be under cover as ladders, especially wooden ones, deteriorate faster when exposed to the effects of weather, particularly extremes of hot and cold, wet and dry. Prolonged contact with water can cause rot. Wooden ladders should be stored in a well-ventilated situation away from radiators or hot pipes which can promote warping. Aluminium alloy ladders should be kept away from wet lime or cement which can corrode. Tie rods, hooks and fastenings are susceptible to corrosion in salt-laden atmospheres, in coastal areas for example.

Marking, registration

Every ladder, step-ladder or trestle should be marked with an identification number and corresponding entry should be made in a card index or register to act as store record and history sheet. Details of inspections, defects and repairs, and issue and return to stores should be noted.

Should a ladder fall or be subjected to overload or shock, it should be carefully inspected and repaired if necessary.

Ladders should be destroyed when they are no longer capable of being effectively repaired and kept in a safe condition and the appropriate entry made in the records.

Inspection

The prime requirements for a ladder are its:

(a) strength;
(b) rigidity (it must not sag, whip or sway too much);
(c) durability;
(d) comfort in use;
(e) resistance to weather;
(f) lightness and portability (except for fixed ladders).

Rearing and extending a ladder

The best angle for a ladder to rest for climbing is 75 degrees from the horizontal, or 4 in 1.

A short step- or rung ladder is easily raised by one man. The foot is placed against a wall or other solid object and the top lifted and 'walked up' hand over on the stiles until the ladder is upright.

To raise a ladder of 10–15 m, it should be laid on the ground with the base at the spot where the ladder has to stand. The heaviest man available should place a foot on the bottom rung and apply his weight, while the other man or men lift. As the ladder is raised the man 'footing' should reach forward and grasp the stiles. Once the ladder is upright it should be carefully eased over to its resting place. It is necessary to have a good sense of balance to do this.

At all times during raising or lowering, a ladder must have adequate weight applied to the base and, if necessary, two men should 'foot' the ladder to prevent it kicking.

A rope-operated extension ladder should be raised to the upright in a closed position and then extended a few rungs at a time.

Securing a ladder

For short ladders and short periods a man standing at the foot of the ladder is acceptable as a minimum precaution. The surest way of preventing foot movement is to lash the ladder to some convenient structure. It could be lashed to a scaffold tube or to ground stakes or anchors, etc. at the base. A top lashing to the two stiles is even better as this also prevents sideways movement. This is essential for ladders over 7 m long where it becomes difficult to control sideways movement. Ladders must never be secured by their rungs.

Ladder safety checklist

1. Do not erect on sloping ground.
2. Do not erect on movable objects.
3. Do not erect in front of a door that may be opened.
4. Do not erect against a slippery surface.
5. Do not erect at a shallow angle.
6. Do not erect horizontally as a plank or bridge.
7. Do not erect at too steep an angle.
8. Do not use tools or do jobs requiring two hands while standing on a ladder.
9. Do not drop materials from a ladder.
10. Do not straddle from the ladder to a nearby foothold.
11. Do not allow more than one person up a ladder at a time.
12. Do not use a ladder which is too short.
13. Do not use a defective ladder.
14. Do not use a makeshift or 'home-made' ladder.
15. Do not overreach.
16. Do not overload a ladder or support it with a rung bearing on a board.
17. Do not slide down a ladder.
18. Do not carry sheets of material, especially if it is windy.
19. Do not carry a ladder while riding a bicycle.
20. Do not use an alloy or wet ladder near electrical conductors.
21. Always place a ladder on a firm level base.
22. Always set at an angle near to 75° from the horizontal, i.e. 4 in 1.
23. Always tie the ladder in position, if possible top and bottom.

24. Always make sure the ladder projects above the climbing-off level.
25. Always make sure soles of boots are clean.
26. Always carry a ladder with end high enough to clear people's heads.
27. Always get help with long ladders.
28. Always report all defects immediately.
29. Always inspect ladders regularly and keep records.
30. Always store ladders carefully.

Ropes

The term rope is used for both fibre and wire ropes. Wire ropes do not fall within the scope of this book. Fibre ropes may be classified into two groups:

    (a) natural fibre ropes;
    (b) man-made fibre ropes.

Natural fibre ropes

This class of rope is made from vegetable fibres obtained from a variety of plants grown in several different countries.

Manila, sisal, hemp and coir are the types that come readily to mind but for the purpose of this chapter just two, manila and sisal, are considered, as these are the ropes most commonly used in the industry.

Manila

This is made from the fibres of a tree which grows in some hot countries, e.g. the Philippines, and produces long, strong fibres.

Manila is brown in colour, and it stands exposure to the weather and damp better than sisal and does not kink so readily. It is reasonably soft and pliable. The best quality manila is stronger than any other kind of natural fibre rope.

BS 2052 identifies one grade only: Grade 1 – a blue thread in two of the strands.

Sisal

Made from the leaves of a cactus-like plant which grows in semi-tropical countries like Kenya, these fibres are not as long or as strong as manila.

Sisal is a light-creamy-coloured fibre and is coarse and rough to handle. It is equal to Grade 2 manila in strength.

Characteristics and care

All ropes are constructed with several fibres twisted together to make a yarn, several yarns are twisted together to make a strand and three or more strands are twisted together into a rope.

The size of ropes is based on either the circumference of the rope or the diameter. Ropes are made in a great variety of sizes, from 7 mm to 144 mm in diameter. However, the two most common sizes in use are 24 mm and 18 mm diameters. Natural fibre ropes are made up in coils of 120 fathoms, which is about 220 m. To uncoil rope the coil should be laid flat with the inside end at the bottom, and the inside end pulled up through the coil.

Because natural fibre ropes are of vegetable origin, they are prone to rotting if they are not properly maintained. The most common cause of rotting is leaving the rope in a damp condition. Ropes are often seen on site hanging from a gin wheel in the rain, with the best part of the rope lying in the mud. It may not be generally known, but rot starts from the *inside* of the rope.

Inspection of a rope by opening it out should show the inside to be clean and dry. If the inside has a smell different from the distinctive 'ropey' smell, the whole of the rope should be inspected. It certainly should not be powdery or discoloured. Ideally, natural fibre ropes should be stored in a condition that will enable a coil to be hung on wooden pegs in a well-aired store kept at a temperature of 10–12°C with the humidity at 40–45 per cent.

If the rope is too wet it must be dried out naturally and not force-dried. Too much heat will make the fibres brittle and useless.

All ropes must be kept well away from acids, active gases, cleaning materials and any other chemicals which may rot the fibres. It is far better to scrap any rope which may have become contaminated than to take a chance with it.

If the condition of the inside of the rope is satisfactory, it should then be inspected from the outside. Ropes can be assessed in three categories:

    (a) *Condition 1 (excellent)*. – This is a new rope, or one with *no* signs of external or internal wear.

    (b) *Condition 2 (good)* – A used rope, but with only slight signs of wear and tear so that there is little loss of strength.

    (c) *Condition 3 (fair)* – A rope which has had an appreciable amount of use, or which is fairly old, and which shows clear indications of chafe or wear internally or externally, or where there is any other deterioration of the fibres.

This rope should not be used for any lifting purpose, but it may be used where failure is unlikely, or where failure would be of little importance and would not cause damage or hazard. Natural fibre ropes may during manufacture be treated with chemicals to prevent rot and water absorption, but this must be specified when purchasing.

Man-made fibre ropes

The first artificial fibre to be used for ropes was nylon. Further developments have now produced ropes made from Terylene, polyethelene, polypropylene and others. All of these materials have some characteristics in common, but each of them has some characteristics of its own.

Comparisons between man-made and natural fibres show that the man-made fibres are stronger, less susceptible to chemical attacks, less able to absorb water, and completely resistant to mildew and rot.

They are non-flammable, but they will melt if heated. Care must be taken to avoid undue friction and when used in conjunction with a gin wheel or rope block, it is essential that the rope size matches the rope groove in the wheel or sheave.

Unnecessary exposure to strong sunlight, heat or chemicals should be avoided. If the man-made ropes do become contaminated, they must be washed thoroughly with clean water, and allowed to dry naturally.

Man-made fibre ropes must be stored as for natural fibre ropes.

4

Putlog scaffold

Putlog scaffolds, sometimes called bricklayers' scaffolds, depend for their support on the walls of the buildings on their inner side, and rows of standards on their outer side.

Components of putlog scaffold

Foundations

It is essential that a good base is provided for the standards. The ground should be level and well consolidated. Timber soleplates at least 228 mm wide and 38 mm thick should be placed as support for the baseplates for the standards. Soleboards should, wherever possible, parallel to the building, and should support at least two standards. There is a very good reason for this, in that the soleboard will ensure that the load carried by each standard is evenly distributed over a larger area than the baseplate, and prevent the standards sinking into the ground.

Standards

Standards should be erected vertically in a row parallel to the wall at intervals of 2 m for a heavy duty putlog, or 2.1 m for a general duty putlog according to the load.

The working platform should be wide enough to accommodate five boards, with a gap left between the boards and the wall to allow for a level or plumb line.

Ledgers

Ledgers are fixed at approximately 1.37 m intervals to the standards with right-angle couplers. A height of 1.37 m is taken as the most convenient height for bricklayers to work to before it becomes necessary to erect a further lift.

Only one working platform should be in use at any one time. Ledgers must remain in position as the scaffold is erected higher but if necessary the first ledger may be removed, providing another ledger is fixed at not more than 1.8 m from from the ground.

Putlogs

Putlogs are fixed with one end to the standard and, at the other end, the blade of the putlog, or adaptor, must rest on the brickwork. The putlog or putlog adaptor must rest 'flat' on the brickwork, to a depth of at least 75 mm (Fig. 4.1).

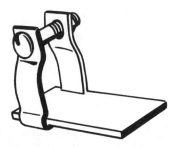

Fig. 4.1 Putlog adaptor

On the working platforms, intermediate putlogs are fixed to the ledgers. It may be necessary from time to time to remove the intermediate putlogs, especially when raising the working lift. If so, it is essential to leave the main putlogs in position, secured to the standards with right-angle couplers.

Bridle

If a putlog is required opposite an opening in the building face, a bridle should be used. A bridle tube is a length of tube underslung from the inner ends of the putlogs on either side of the opening with right-angle couplers. The intermediate putlog will now be supported by the ledger and bridle tube.

Bridle tubes should be supported by at least two putlogs on each side of the opening (Fig. 4.2).

43

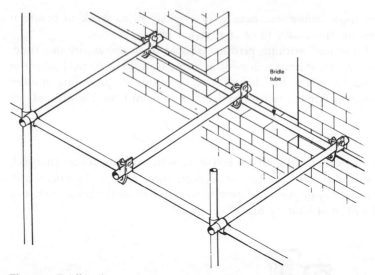

Fig. 4.2 Bridle tube

Guardrails

Guardrails must be fixed at a height between 914 mm and 1.15 m and toe-boards must be provided. The space between the guardrails and toe-boards must not be greater than 0.75 m.

It is important to remember that materials should not be stacked higher than the toe-boards, unless brick-guards or similar are fitted.

Ties

The security of the scaffold will depend on the effectiveness of the ties.

With a putlog scaffold all ties must be two-way ties.

Bracing

The only form of bracing necessary, is longitudinal or sway bracing, providing the brace is fixed from the base to the full height of the scaffold and adequately tied.

Putlog scaffolds (using GKN Mills fittings)

Sequence of erection

In Chapters 4–10, I will be describing what I call a 'sequence of erection'; that is to say, a method by which a scaffolder may set about the task of erecting a particular scaffold. No matter how experienced a scaffolder may be at erecting scaffolds, there are bound to be some which he has not tackled before.

The methods of erecting scaffolds are explained clearly, with logical step-by-step instructions, and easy-to-follow diagrammatic sketches.

1. Measure and position soleboards, set out baseplates on soleboards (Fig. 4.3 and 4.4).
2. Fix putlog adaptors to ends of transoms and measure putlogs for board width. Place putlog flat on finished brickwork.
3. Fix right-angle coupler to putlog. Fix first standard to putlog, level and fix (Fig. 4.5).
4. Repeat operation to second standard.
5. Fix 2 right-angle couplers on first and second standards, fix ledgers in couplers (Fig. 4.6).

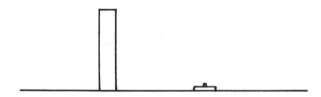

Fig. 4.3

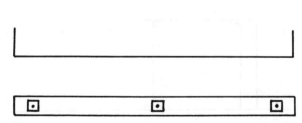

Fig. 4.4

45

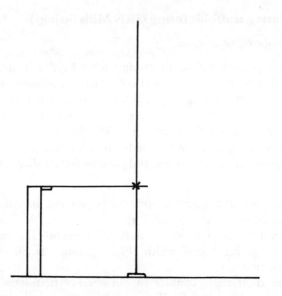

Fig. 4.5

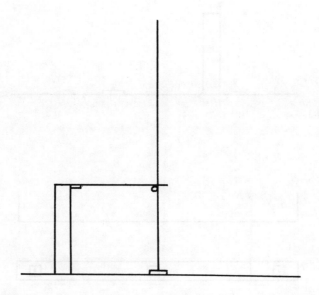

Fig. 4.6

46

6. Level and fix a foot-tie ledger approximately 150 mm from baseplate (Fig. 4.7).
7. Fix bridle tube 100 mm approximately from wall (Fig. 4.8).

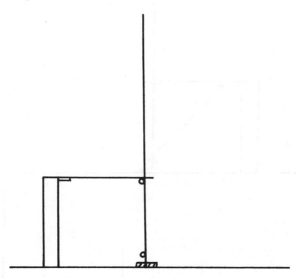

Fig. 4.7

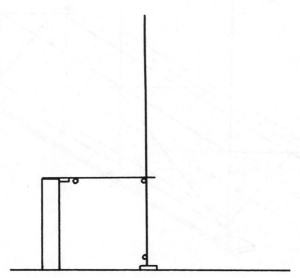

Fig. 4.8

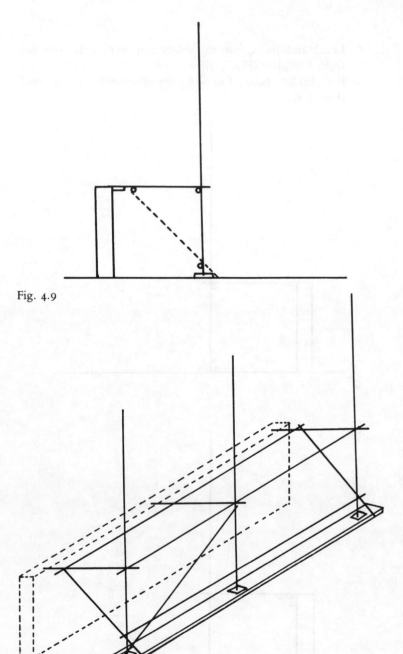

Fig. 4.9

Fig. 4.10

Figs. 4.3–4.10 Sequence of erection for a putlog scaffold

48

8. Fix right-angle couplers on bridle tube and foot-tie ledger. Place braces, plumb standards and secure braces, secure standards (Fig. 4.9).

9. Fix intermediate standard and transom. Fix sway brace, plumb standards and secure (Fig. 4.10).

Safety checklist (putlog scaffold)

1. Is the base sound, level and well-consolidated?
2. Are the soleplates at least 228.6 mm wide and 38 mm thick?
3. Are there at least two standards per soleboard?
4. Are standards pitched on steel baseplates?
5. Are the standards plumb?
6. Are the standards set between 1.82 m and 2.43 m apart?
7. Are the joints in both the standards and ledgers staggered and those in the ledgers as close to the standards as possible?
8. Are the ledgers connected to the standards with right-angle couplers?
9. Is the spade end of a putlog laid flat on the brickwork to a depth of at least 76 mm?
10. Are the main putlogs within 300 mm of the standards?
11. Is the bridle tube secured with right-angle couplers?
12. Has a sway brace been fixed from the base to the full height of the scaffold?
13. Has the scaffold been properly secured to the structure with sufficient ties?
14. Is the working platform close-boarded and evenly supported?
15. Is the scaffold wide enough to accommodate whatever materials need to be stacked?
16. Have guardrails and toe-boards been properly fixed?
17. Have brick-guards been fixed?
18. Are all ladders, used for access, sound and free from defects?
19. Are all ladders correctly positioned at an angle of 4 in 1?
20. Are all ladders properly secured?
21. Has the scaffold been overloaded?

5

Independent tied scaffolds

An independent tied scaffold consists of two rows of standards, each row parallel to the building. The inner row of standards is set as close to the building as practical, or not further away than just enough to allow for an inside board between the inside standards and the building or structure. The distance between the inside standards and the outside standards will be governed by the number of boards which will be required.

Ledgers are fixed to the standards with right-angle couplers and the ledgers, like the standards, are fixed parallel to the building. Transoms are fixed to the ledgers with putlog couplers and the transoms are fixed at right angles to the ledgers. Braces are fixed diagonally to the ledgers or standards. Sway bracing or longitudinal bracing is fixed to the standards or transoms and is fixed across the face of the scaffold.

The self-weight of the scaffold together with any loads on it are transferred to the ground via the standards.

The type of load, whether it is a distributed load or a point load of any other loading, may be specifically designed.

If no load-rating is quoted by the specification then one should be selected from the Code of Practice table of loads.

The spacing of the standards or the bay length depends on the height and loading of the scaffold. The spacing of the ledgers or the lift height is normally 2 m but in certain circumstances lifts may be greater, provided the standards are capable of supporting the load.

Types of independent tied scaffold

It is generally accepted that there are five types of independent tied scaffold:

   (a) inspection and very light duty;
   (b) light duty;

(c) general purpose;
(d) heavy duty;
(e) masonry or special duty.

Inspection and very light duty

Inspection and very light-duty scaffold, as the name suggests, should be used for work of a very light nature and is used mainly for inspections, painting, stone cleaning without heavy tools; it should have a maximum of one working platform only.

It should be no less than 3 boards wide, have a maximum bay length of 2.7 m and a maximum loading of 75 kg/m^2 on the one platform.

Light duty

Light-duty scaffolding is also used for light work, i.e. painting, stone cleaning, glazing, pointing and plastering. It should have a maximum of 2 working platforms, 4 boards wide, a maximum bay length of 2.4 m and a maximum loading of 150 kg/m^2.

General purpose

General-purpose scaffold is, as the name suggests, a general, all-purpose scaffold used by all trades, e.g. brickwork, window fixing, mullion fixing, rendering, plastering. It should have a maximum of 2 working platforms with a maximum loading of 200 kg/m^2 plus 1 working platform with a maximum loading of 75 kg/m^2. It should have a maximum width of 5 boards or 4 boards plus 1 inside board, and a maximum bay length of 2.1 m.

Heavy duty

Heavy-duty scaffold is used for heavier loads, e.g. brickwork, blockwork, heavy cladding, etc. It should have a maximum of 2 working platforms with a maximum loading of 250 kg/m^2, plus 1 inspection platform with a maximum of 75 kg/m^2.

It should have a maximum of 5 boards or 5 boards plus 1 inside board, or 4 boards plus 1 inside board, and a maximum bay length of 2 m.

Masonry or special duty

Masonry/special-duty scaffold is used for masonry, blockwork or very heavy cladding.

It should have only 1 working platform with a maximum loading of 300 kg/m² plus 1 inspection platform with a maximum loading of 75 kg/m². It may have a 6 or 8-board wide platform, but must be restricted to a maximum bay length of 1.8 m.

Independent scaffold (using GKN Mills fittings)

Sequence of erection

1. Measure and position soleboards. Set out baseplates on soleboards (Fig. 5.1).
2. Select 4 tubes. 2 of which are to be used as ledgers and 2 to be used as transoms. Measure and fix 2 right-angle couplers to each ledger and transom (Fig. 5.2).

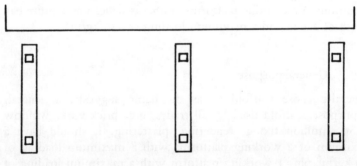

Fig. 5.1

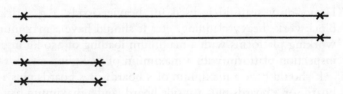

Fig. 5.2

3. Erect first standard, fix ledger to standard approximately 150 mm from the ground. Fix transom to standard above the ledger (Fig. 5.3).
4. Fix second standard to transom, level transom and fix (Fig. 5.4).
5. Fix second ledger to standard (Fig. 5.5).
6. Fix third standard to ledger, level ledger and fix (Fig. 5.6).
7. Fix second transom to standard (Fig. 5.7).
8. Fix fourth standard, level ledger and fix, fix transom (Fig. 5.8).

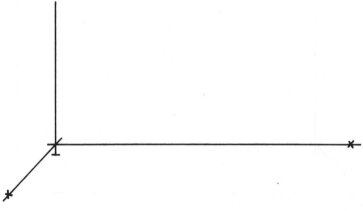

Fig. 5.3

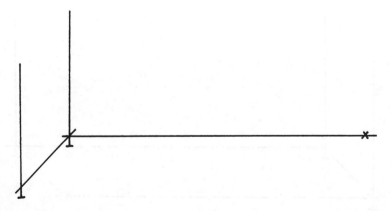

Fig. 5.4

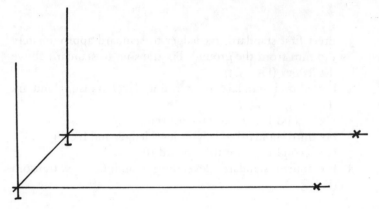

Fig. 5.5

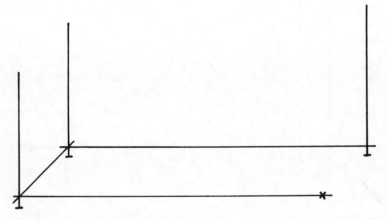

Fig. 5.6

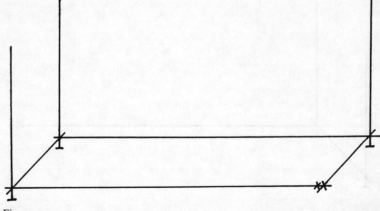

Fig. 5.7

54

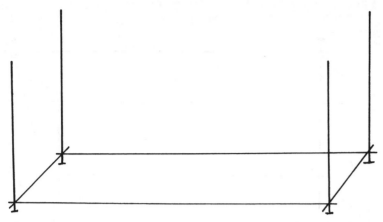

Fig. 5.8

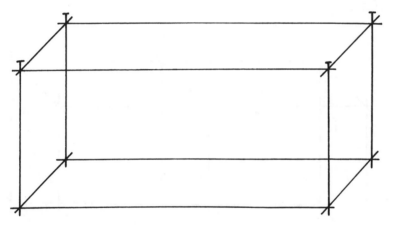

Fig. 5.9

9. Measure 1.8 m from ledgers, fix 2 right-angle couplers to each standard, 1 right-angle coupler positioned to receive ledger, the other right-angle coupler to receive transom. Place ledgers and transoms in couplers (Fig. 5.9).
10. Position right-angle couplers to receive braces. Place braces in couplers, plumb standards and secure them. (Fig. 5.10).
11. Measure and fix intermediate standards, place and fix intermediate transoms (Fig. 5.11).
12. Fix sway brace, plumb standards and secure (Fig. 5.12).

Fig. 5.10

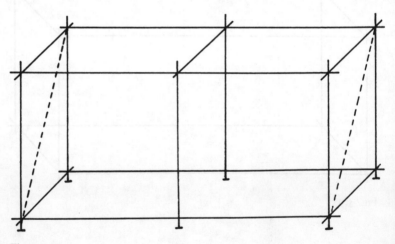

Fig. 5.11

Checklist (independent scaffolds)

1. Are the standards plumb, and supported on a firm foundation?
2. Are soleboards required to spread the load?
3. Are ledgers and transoms level?
4. Are the ledgers and transoms correctly spaced?

56

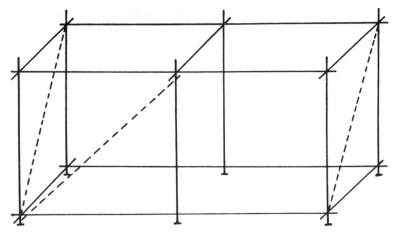

Fig. 5.12

Figs 5.1–5.12 Sequence of erection for an independent scaffold

Fig. 5.13 Independent scaffold using GKN Mills fittings

5. Are the joints staggered both vertically and horizontally in standards and ledgers?
6. Are there sufficient braces?
7. Are the correct number of ties fixed?
8. Are the boards sound and free from defects?
9. Are the boards properly supported?
10. Are the guardrails and toe-boards in place?
11. Are the ladders sound and correctly fixed?
12. Are the correct fittings (couplers) used, and are they properly secured?

6

Birdcage scaffold

The birdcage scaffold is so called because of its appearance. It is normally an internal scaffold and is mainly used for ceiling work in churches, large halls, etc.

It consists of an arrangement of standards with ledgers and transoms supporting a closely boarded platform at the required level. The side and end bays may also be required to form access scaffolding to the walls supporting the soffit.

Birdcage scaffold is the general term used, but it can be divided into two types:

(a) birdcages with more than one lift;
(b) single-lift birdcages.

Birdcage with more than one lift

This type of scaffold is a light-duty scaffold, therefore light loads only should be placed on the scaffold; a maximum of 75 kg/m^2 should be used as a guide, with a maximum standard spacing of 2.5 m in either direction.

It is recommended that, apart from the working lifts on the end or outside bays, a birdcage scaffold has only one working lift.

The foundations for a birdcage scaffold must be pitched on steel baseplates, but may require special consideration for different conditions, e.g. if the birdcage has to be erected on a slope, wedge-shaped pieces of timber should be used up to a maximum of 1 in 4 and in some cases it may be necessary to put foot-ties in. In fact it is a good idea to put in as many foot-ties as possible.

Provision must be made for scaffold that is placed on highly polished wood, mosaic, marble or similar floors. Protective material should be laid under the soleboards, e.g. tarpaulin sheets, hardboard, etc.

Baseplates should be nailed or screwed to the soleboards to prevent movement at the base of the scaffold.

Bearing in mind the type of scaffold used, the first lift height should not exceed 2 m and the subsequent lifts should be at 2 m intervals.

Ledgers and transoms should be fixed to the standards except, if necessary, at the top lift where transoms may be fixed direct to the ledgers.

Sufficient diagonal bracing is required to ensure stability, i.e. for every alternate pair of standards from the base to the full height of the scaffold in both directions. Birdcage scaffolds must be securely tied to columns or the side walls throughout their length and height to prevent movement of the scaffold.

Although it is becoming increasingly popular to leave bracing out of this type of scaffold and extend the ledgers and transoms to butt the walls, great care must be taken to ensure that the scaffold has the facility of internal columns to tie to or at least reveal ties fixed to alcoves or returns. If ledgers and transoms butt the walls plastic tube-caps must be fitted on to the ends of the tubes.

Guardrails and toe-boards are necessary when the working platform finishes more than 155 mm form the walls or is higher than 2 m from the ground.

Single-lift access birdcage scaffold

At first glance a single-lift birdcage may look stable; in fact it may be considerably less stable than a multi-lift birdcage. Because it is one lift, some think bracing may be omitted.

The following are required to ensure stability:

(a) Bracing must be fixed to each corner at least, and every alternate pair of standards in both directions.

(b) There must be foot-ties all the way round and internal standards should be foot-tied in pairs in one direction at least.

(c) Foot-ties must be fixed to the bottom of the standards to which the bracing is connected.

Needless to say, where possible the scaffold should be box-tied round existing columns and if possible should butt the walls with extended transoms fitted with plastic tube-caps to prevent damage to walls.

One final point – it is necessary to stagger the joints in ledgers and transoms. Sleeve couplers are used for this.

Birdcage scaffold

Sequence of erection

1. Measure and position soleboards if required. Set out baseplates.
2. Measure and mark off 2 ledgers and 2 transoms. Fix 2 right-angle couplers on each tube.
3. Erect a standard, fix ledger and transom to standard approximately 150 mm from baseplate (Fig. 6.1).
4. Fix second standard to transom, level transom (Fig. 6.2).
5. Fix second ledger to standard underneath transom (Fig. 6.3).
6. Fix third standard to ledger and level ledger (Fig. 6.4).
7. Fix second transom to standard on top of ledger (Fig. 6.5).
8. Erect fourth standard, level ledger and transom and secure (Fig. 6.6).
9. Measure 1.8 m from foot-tie ledger and mark all 4 standards. Fix 2 right-angle couplers to each standard. Place 2 ledgers and 2 transoms in couplers (Fig. 6.7).
10. Measure and erect 4 intermediate standards (Fig. 6.8).
11. Fix sway brace from foot-tie lift to first lift. Plumb and secure standards. Repeat bracing on all 4 sides (Fig. 6.9). Fix intermediate ledger and transom.
12. Fix centre intermediate standard (Fig. 6.10).
13. Fix intermediate transoms for board support (Fig. 6.11).
14. Deck out platform.
15. Measure 1.8 m up from ledger, mark all standards. Fix 2 right-angle couplers to each standard. Place ledgers and transoms in couplers (Fig. 6.12).
16. Fix sway bracing on all 4 sides, plumb and fix standards (Fig. 6.13).
17. Fix intermediate transoms for board support and deck out platform.
18. Remove foot-tie lift if necessary in places where through access is required but not more than one-half of the bays should be without a foot-tie in a single-lift birdcage.

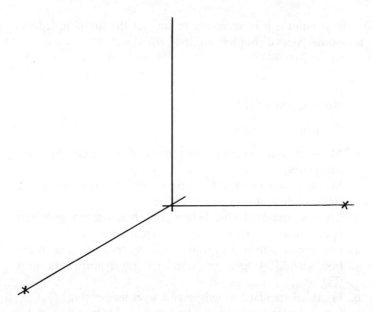

Fig. 6.1

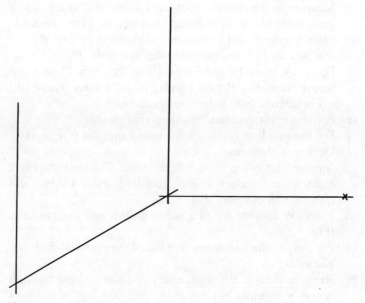

Fig. 6.2

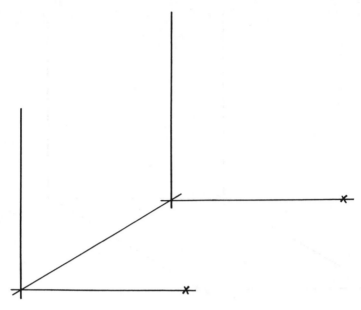

Fig. 6.3

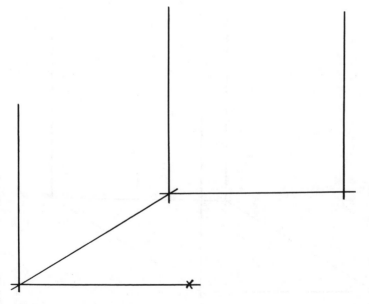

Fig. 6.4

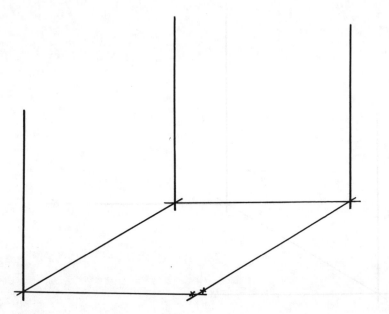

Fig. 6.5

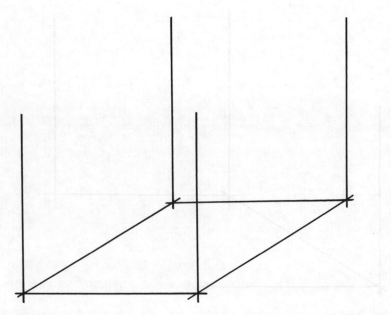

Fig. 6.6

64

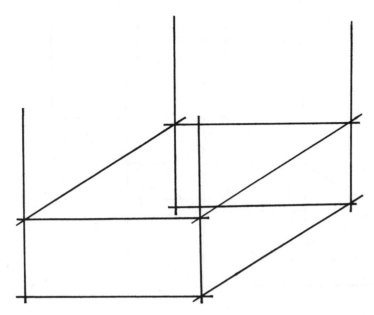

Fig. 6.7

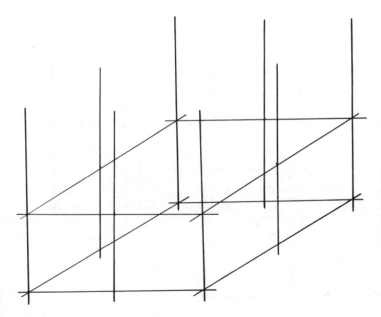

Fig. 6.8

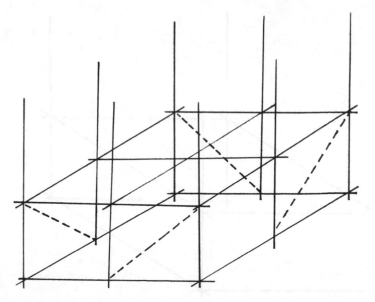

Fig. 6.9

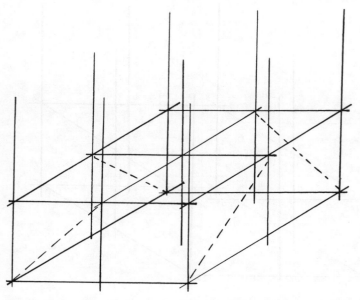

Fig. 6.10

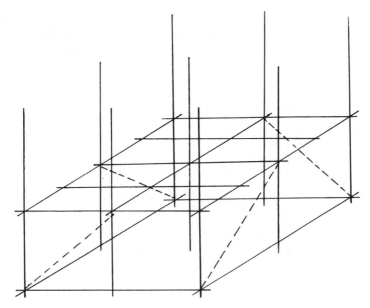

Fig. 6.11

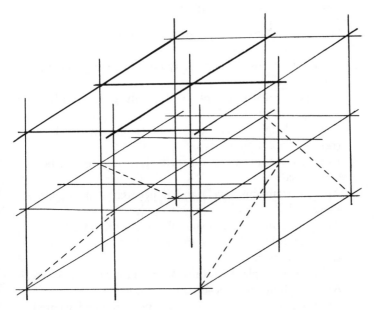

Fig. 6.12

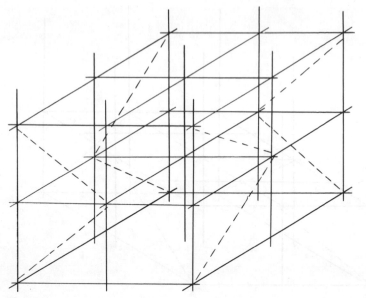

Fig. 6.13

Figs 6.1–6.13 Sequence of erection for a birdcage scaffold

Checklist (birdcage scaffold)

1. Is the floor strong enough to carry the weight of the scaffold and its load?
2. Have soleboards been correctly positioned, e.g. set at right angles to joists?
3. Are the standards plumb, and evenly spaced within the loading requirements?
4. Are the ledgers level and fixed to the standards with right-angle couplers?
5. Are the transoms level and correctly spaced for the boards being used?
6. Are the transoms level, and if necessary, are the joints staggered?
7. Are there sufficient and suitable braces provided?
8. Is the scaffold tied, or are the ledgers and transoms butted against the walls?
9. Is the working platform fully boarded over the whole area?
10. Are guardrails and toe-boards provided?
11. Are the ladders sound, properly supported and secured?
12. Has the scaffold been overloaded?

Fig. 6.14 Birdcage scaffold

7

Truss-out scaffold

A truss-out scaffold is an independent tied scaffold not erected from the ground but supported by a truss-out scaffolding structure projecting from the face of a building or structure.

It is essential that assurances are obtained concerning the ability of the building to support the scaffolding safely.

Only right-angle couplers should be used in the construction of the truss-out.

The standards inside the building from which the truss-out is fixed must be strutted between floor and ceiling and firmly secured to prevent displacement – this part of the truss-out is usually referred to as the 'horse'.

Timber soleplates and headplates must be used to distribute the load. Tubes projecting from the built-up inside scaffold (the horse) are known as needle transoms and must be secured with right-angle couplers and when possible rest on sills and be right up against reveals.

The tie tubes must always be fixed at the back of a window or opening with right-angle couplers.

The inner and outer ledgers should be fixed to and on top of the needle transoms, with right-angle couplers.

Rakers should be set at an angle of not more than 35° from the vertical and be fixed with right-angle couplers with a check coupler fixed immediately underneath and in contact with the ledger coupler. The raker should be fixed to the outside ledger with a right-angle coupler and the lower end of the raker secured to prevent displacement. The upper end of the raker should be fixed as close to the needle transom as possible.

The unbraced length of rakers should never exceed 3 m.

The standards for extending the scaffold vertically are puncheoned off the bottom ledger next to the rakers with right-angle couplers, with a check coupler fixed on top.

The scaffold should be erected in accordance with the same recommendations as access scaffolds.

The maximum height for a truss-out scaffold is 12 m.

The first lift above the truss-out should be tied back to the building. The ties at higher levels should be distributed at the same frequency as for a ground-based independent tied scaffold.

Truss-out scaffold

Sequence of erection

1. Place soleboards in position. Place baseplates on soleboards.
2. Measure and place 2 ledgers and 2 transoms in position (Fig. 7.1).
3. Place adjustable jacks into tops of standards. Place standards on baseplates. Secure ledgers and transoms in place (Fig. 7.2).
4. Measure and secure first lift ledger. Measure and secure second foot-tie ledger (Fig. 7.3).
5. Place braces in position, fixed from inside foot-tie ledger to first lift ledger. Plumb standard (Fig. 7.4).
6. Place adjustable jacks into tops of standards. Place standards on baseplates. Secure ledgers to standards. Secure transoms to standards. Plumb standards (Fig. 7.5).
7. Measure and fix ledgers for needle tubes. Place needle tubes on top of inside ledger, underneath outside ledger and secured to standards with right-angle couplers. Fix second lift braces and plumb standards. Fix second lift ledgers and

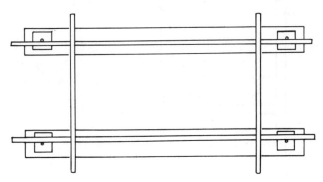

Fig. 7.1

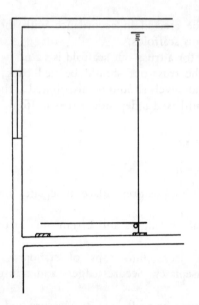

Fig. 7.2

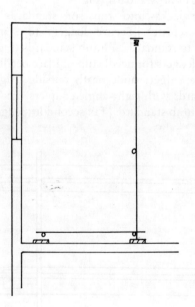

Fig. 7.3

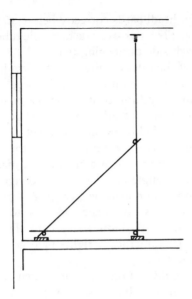

Fig. 7.4

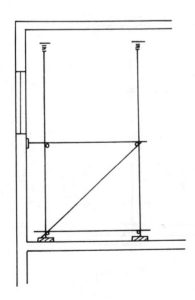

Fig. 7.5

transsoms. Position headboards correctly before adjustable jacks are tightened. Fix a two-way tie to needle tubes with one tie tube on each side of opening (Fig. 7.6).

The number of boards required in the width will determine the measurements needed on the needle tubes. Allowing 300 mm to accommodate an inside board, and assuming a 4-board wide scaffold is needed it will be necessary to make the following measurements.

8. From the building measure 300 mm and mark the needles; from this mark measure 1.0 m and put another mark. On these marks fix right-angle couplers and into right-angle couplers place inside and outside ledgers (Fig. 7.7).

9. Fix right-angle couplers to outside ledgers to receive rakers. Place rakers in position and secure. Fix check coupler (right-angle) underneath right-angle coupler on outside ledger (Fig. 7.8).

10. Bottom of raker should rest on some convenient window ledge or feature of the building. To prevent rakers becoming dislodged, buckling or splaying, lace together with ledgers. It may also be necessary in certain

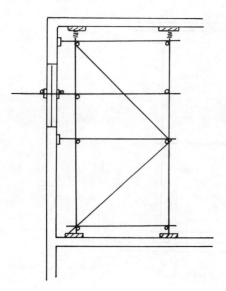

Fig. 7.6

74

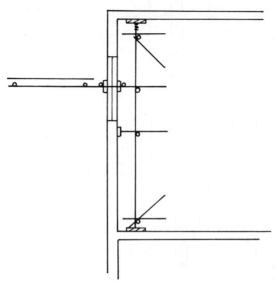

Fig. 7.7

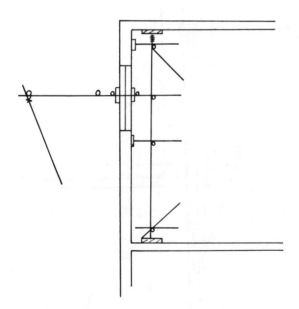

Fig. 7.8

75

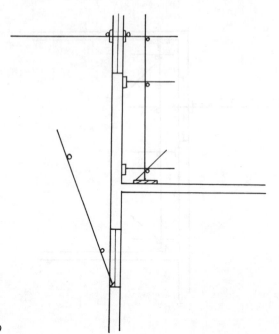

Fig. 7.9

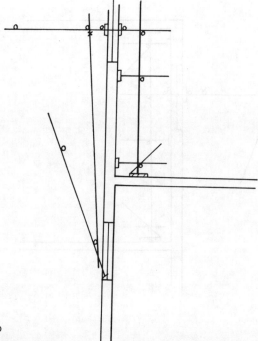

Fig. 7.10

circumstances to use the lacing ledger to provide additional support for the scaffold (Fig. 7.9).

11. Figure 7.10 shows an inside supporting raker with a safety check coupler underneath the ledger coupler.

12. Figure 7.11 shows an intermediate raker fixed from the intermediate lacing ledger to the inside ledger and a short transom for additional support.

13. The build-up scaffold can now be erected. Standards or puncheon tubes must be placed into right-angle couplers fixed to the ledgers and have the security of a check coupler fixed immediately on top of the right-angle coupler fixed to the ledger (Fig. 7.12).

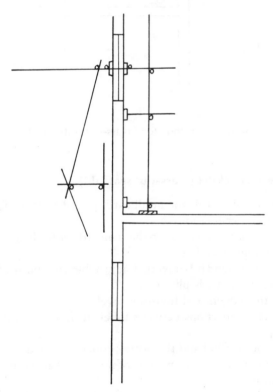

Fig. 7.11

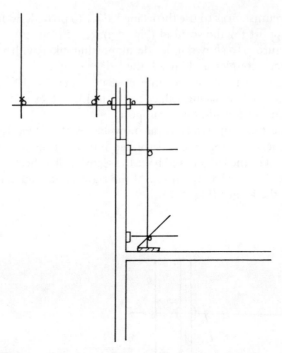

Fig. 7.12

Figs 7.1–7.12 Sequence of erection for a truss-out scaffold

Safety checklist (truss-out scaffold)

1. Are the soleboards and headboards placed at right angles to floor and ceiling joists?
2. Are the needles secured to the standards or the ledgers with right-angle couplers?
3. Have the standards inverted adjustable jack on top?
4. Are the standards plumb?
5. Are the ledgers and transoms level?
6. Are all connections between tubes made with right-angle couplers?
7. Are the scaffold and the horse properly braced?
8. Are the rakers set at an angle of no more than 35° from the vertical?
9. Are the tops of the rakers fixed to the ledgers with right-angle couplers, and are they supplemented with check couplers?

78

10. Are the bottoms of the rakers supported on a firm ledge and tied into it?
11. Are the rakers longer than 3 m?
12. Are the rakers fixed to every standard?
13. Are the puncheons fixed to the ledgers and the transoms to the ledgers with right-angle couplers, and are they supplemented by check couplers?
14. Are all joints made with sleeve couplers and are they staggered?
15. Is there a sufficient number of through ties especially at the first lift above the truss-out?
16. Is the working platform close-boarded and evenly supported?
17. Have the guardrails and toe-boards been properly fixed?

Fig. 7.13 Truss-out scaffold

8

Suspended scaffolds

Most people think of suspended scaffolds as cradles and there is some confusion between these terms, so they are worth defining.

A cradle is a lightweight scaffold suspended by a fibre rope and pulley block or winch and is used for work of a light nature. The platform is at least 440 mm wide, and any fibre ropes used with pulley blocks are not more than 3 m apart. For simplicity this type of scaffold will be referred to as a manually operated suspended scaffold.

The term 'suspended scaffold' has come to mean a power-operated suspended scaffold which is beyond the scope of this book. Two or more platforms are used, suspended one above the other, to meet the need for faster building programmes where more than one working level is needed to follow on with finishings.

A good example of this type of suspended scaffold was the aluminium suspended access system used on the National Westminster Bank Tower in London. Over 300 m of articulated suspended scaffolding encircled its curved and irregular perimeter.

Suspended scaffolds are most frequently used for painting, glazing, cladding, scaling, cleaning and any light work on tall buildings or structures above busy streets, or where other obstructions intervene to make it neither feasible nor economic to erect scaffolding from the ground.

There are three major factors in considering the use of suspended scaffolds in preference to traditional scaffolding whether it is tubular or a proprietary system. First, economy: any contractor worth his salt will always compare the cost of one system against the cost of another system.

The second factor to be considered is the versatility of the system. Suspended scaffolds may be traversed, raised and lowered, height being virtually irrelevant. Sections can be built up to span the perimeter of a structure and stagings can be adapted to meet specific requirements, e.g. two and three-tier stagings for cladding work, underside stagings for underneath hopper heads, sloping stagings for lighthouses, dam faces, etc.

The third factor is the speed of erection and, as buildings or structures increase in height, the sheer physical effort of erecting traditional scaffolds. Labour charges and hire charges for traditional scaffolds for high-rise structures can never undercut labour cost and hire charges for power-operated suspended scaffolds.

A variety of hand-operated winches is available for raising and lowering manually operated suspended scaffolds, but generally speaking there are two main types:

(a) drum winch, with single or double handles, e.g. Strateline or LAHO;
(b) pump-action winch, such as Tirfor.

The Construction (Working Places) Regulations 1966

These govern the use of all types of suspended scaffold as follows:

Regulation 19(i) states that the Construction (Lifting Operations) Regulations apply to all ropes, lifting tackle and lifting appliances, particularly paragraphs 3–14.

Paragraph 3. Ropes, winches or other lifting appliances should be correctly rigged. Outriggers, joists, runways must be safely anchored.

Paragraph 4. Winches or other lifting appliances shall have:

(a) brakes which apply when the operating lever is released;
(b) protection from the weather, dust or material likely to cause damage.

Paragraph 5. Outriggers should be of adequate length and strength, be horizontal and correctly spaced, have adequate stops at their outer ends.

Paragraph 6. Counterweights must be securely attached to the outriggers, at least three times the overturning moment or load.

Paragraph 7. Platforms must be hung clear of the building or face of the structure.

Paragraph 8. Every runway, joist or rail-track should be:

(a) strong enough;
(b) free from defects;
(c) have stops at each end;
(d) be bolted or adequately secured to its supports.

Paragraph 9. Suspension ropes must be:

(a) properly secured to the outriggers or other supports and to the platform framework;
(b) kept in tension.

Paragraph 10:

(a) Winches must have at least two turns on the drum when the platform is at its lowest position.
(b) The length of the rope must be clearly marked on its winch.

Paragraph 11. Platforms must be prevented from undue tipping, tilting or swinging whilst in use.

Paragraph 12. Steel wire rope must be used for suspension of all platforms except lightweight cradles where fibre ropes and pulley blocks may be used.

Paragraph 13. Platforms must be close-boarded:

(a) At least 400 mm wide on lightweight cradles.
(b) 640 mm wide on all other types, if used only as a footing.
(c) 870 mm wide if used for the deposit of materials.
(d) They must never be used to carry another higher platform.
(e) They must be as close as possible to the face of the building, but where because of the nature of their work men sit on the edge of the platform, the distance between the platform and the building may be a maximum of 300 mm.

Paragraph 14. If a suspended scaffold is carried on fibre ropes and pulley blocks they should be not more than 3 m apart.

Ropes

Care should be taken with the choice of rope if fibre ropes are used for operations involving aggressive chemicals (as in building cleaning).

The fall rope should be of superior quality or Grade 1 manila, at

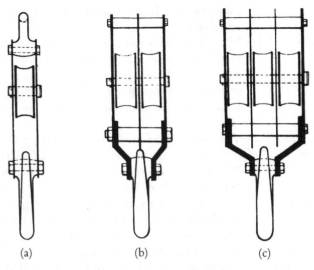

Fig. 8.1 Pulley blocks (a) single (b) double (c) triple

least 57 mm in circumference and should be inspected before every use.

Fibre rope used for lightweight cradles must not have any splices.

Pulley blocks must have a safe working load of 227 kg.

Manually operated lightweight cradles

Manually operated lightweight cradles are working platforms other than boatswains chairs, suspended by fibre ropes and operated solely by hand, or by steel wire rope and operated by manually powered machine.

They are capable of being raised and lowered vertically and also capable of travelling horizontally. More definitions are required at this point.

Definition of terms – suspended scaffolding

Climbing   A lifting machine through which a wire rope
device   passes, controlled either by friction grips or by
              turns of the rope round drums within the

	machine, the lower end of the rope not being anchored.
Cradle	The complete assembly, including the working platform, toe-boards, guardrails, and stirrups.
Outrigger	The assembly of beams, poles, joists, tubular scaffold framework or proprietary brackets to which the upper ends of the suspension members are secured.
Projection length	The length of the portion of the outrigger between the fulcrum point and the point of suspension. Where there are two suspension points, midway between the two.
Stirrups	The end frames supporting the platform and to which the blocks or lift machines are attached.
Suspension member	The assembly of rope and shackles joining the platform unit to the outrigger.
Suspension points	Where the suspension gear is connected to the cradle stirrup at the bottom or the roof rig at the top.
Suspension rope	The rope passing through the pulley block or suspension shackles, commonly called the 'fall rope'.
Tailing length	The lengths of the portion of the outrigger between the fulcrum point and the rear anchorage or centre of gravity of the counterweight.
Winch	A lifting appliance in which the hoist rope is anchored to and reeled on to a drum.
Working platform unit	A single unit consisting of the framework and decking.

Types of cradle

There are many types of manually operated cradle available for use, constructed in metal, wood, fibreglass, etc., but there are two general types: 'one-man' for light-duty work; and 'two-man' for general duties.

Types of suspension

Travelling	The cradle may travel horizontally as well as vertically.
Fixed	The cradle can be moved only vertically.

Built-in	Cradles with permanent tracks built in to the structure, but these are usually powered by electricity.

Outriggers

There are three main types of outrigger used: timber poles, scaffold tubes and rolled sections.

Timber poles

Timber poles are the traditional outriggers for light-duty work such as industrial painting or building cleaning. Larch and Norway Spruce are the timbers mainly used, and they have a maximum length of 6.5 m with a natural taper.

The thick butt end should be at least 115 mm in diameter and this end must always be the overhang, to give the outrigger the

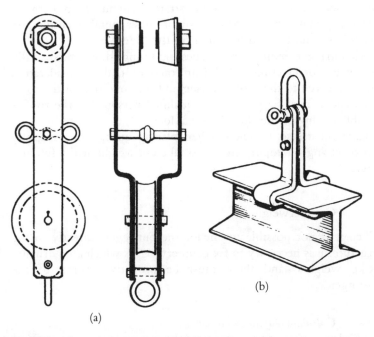

(a)

(b)

Fig. 8.2 RSJ track for travelling horizontally, and jockey roller which runs on lower flanges. (a) Joist roller (jockey) (b) Joist-clip and shackle

maximum available strength. The tapered end should not be less than 65 mm in diameter.

The projection length for a single outrigger should never be more than 450 mm. Timber poles may be tied down with a square lashing with 6 mm diameter wire lashing. The track may also be tied to the outrigger, with 6 mm diameter rope. Wire ropes should not be subjected to more than one-sixth of their breaking load, to ensure a margin of safety of 6 : 1, and lashings should have at least three turns.

Usually three outriggers will be needed for each section of track, depending on the total weight to be carried.

Each track should have an outrigger no more than 450 mm from its end (Fig. 8.2).

Scaffold tubes and rolled sections

Scaffold tubes and rolled steel section outriggers are fixed in the same way as timber poles, but care must be taken to protect the wire ropes from acute bending strains around sharp corners. Timber packing pieces or hessian or other suitable cushioning material should be used to round off the edges.

All outriggers built up from tubes and fittings should be specially designed. For erecting a 'build-up' on a roof, due consideration must be given to the roof covering and the condition of the roof. Assurances must be given by a qualified person that the roof is capable of supporting the additional load.

Care must also be taken when placing outriggers, especially when relying on coping stones to take the weight at the fulcrum point.

Counterweight

When it is not possible to anchor the outriggers directly to the structure, it is necessary to fix counterweights which are normally 25 kg weights, and these must be properly secured to the outrigger.

Calculation of the counterweight

To find the amount of counterweight required, it is necessary to know the weights of cradles and their contents.

The following example shows how this may be arrived at.

Example. A 2 m fixed cradle for 2 men:

The self-weight is:

	kg
Cradle	61
2 fall ropes at 150 m	90
4 blocks	13
Total on 2 ropes	164

The imposed loading for 2 men who may be near one end of the platform:

	kg
1 man at end	75
Small tools	5
	80

Second man further from end $\frac{3}{4} \times 80$ kg	60
Total weight on 1 rope	140

kg

Total weight at one end or other = 82 kg + 140 kg = 222
(i.e. half the total load on 2 ropes)

Add an impact factor of 10 per cent	23
Total suspended load =	245

Formula for amount of counterweight to be used.

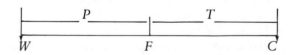

Where W = weight of cradles and contents
C = counterweight without factor of safety
P = projection
T = tailing length
F = fulcrum
$$\text{Then } C = \frac{W \times P}{T}$$

To this a factor of safety of 3 must be applied so that the dead weight to be applied at the end of the tailing length of the outrigger is:

$$3C \text{ or } 3 \times \frac{W \times P}{T}$$

To take this calculation stage by stage:
The maximum load is 245 kg.
Multiply this by a projecting length of say 0.45 m

$$245 \times 0.45 = 110.25$$

Divide this figure by the tailing length of say 7 m

$$110.25/7 = 15.75$$

The balancing weight is therefore 15.75 kg.
This has to be multiplied by 3 to obtain a safety factor of 3

$$15.75 \times 3 = 47.25$$

The counterweight required is therefore 47.25 kg or 2×25 kg weights fixed on the tailing length of each outrigger.

Note: The counterweight calculation is an assessment for **each outrigger** (not the total).

Also, when two or more cradles are suspended from build-up, and there is any chance that the weight of both cradles will be imposed on any one point in the track at the same time, then the counterweight calculation must take this into account.

Typical weights (kg)

	kg
2 m cradle complete with stirrups	61
150 m × 57 mm rope	45
Tirfor wire rope	7
Tirfor T.M. cradle winch	6
Strateline cradle winch	26
Blocks (each)	3
Jockey	5
Girder clip or hanging shackle	5
Traversing line	2
Stop end	2
3 m rolled steel joist	45
3.5 m platform	25

88

	kg
Toe-board	5
One man, normal build	76
Painters' tools and materials, per man	2
Impact loads, plus 10 per cent weight of men and tools	8

Extra allowances must be assessed for any special or extra materials in the cradle or attached to it, or being lifted up into it.

9

Slung scaffold

It is said by many scaffolders that a slung scaffold provides the greatest challenge to their skill. This is true to some extent, but provided the erection of a slung scaffold is done to a logical sequence the average scaffolder who has never erected one before, will find it less hazardous than he first thought.

A slung scaffold is suspended at a fixed height either below load-bearing projecting brackets or beams, or from the structural members of a roof or other overhead structure. The suspension may be by tubular members or by lifting gear and wire ropes which are not provided with the means of raising or lowering a suspended scaffold.

Because of the nature of its uses and loading requirements, a slung scaffold should be specially designed.

Wire rope slung scaffold

The purpose of a wire rope slung scaffold is to provide a working platform for the underside of a structure without restricting movement in the area below. Common applications are for gaining access to the ceilings of large halls, cinemas, bridge soffits, oil rigs, etc.

Suspension ropes

It is becoming increasingly popular to use 12 mm high tensile, flexible wire rope with 75 mm soft eye splices in preference to the 9 mm wire rope. Although the safe working load of the wire rope will be specified, it should have a safety factor of six.

The ropes must not be fouled or weakened by any part of the structure. This problem can usually be overcome by the use of suitable packing around the suspension points. The suspension ropes should be secured to the structure with two round turns, and the dead end and the live ends secured with at least two bulldog grips. In some cases it may be necessary to make suitable anchorage points with tubular scaffolding or prefabricated beams.

Suspension points

Spacings should be limited to 2.4 m centres, or if the spacings have to be greater than this it is necessary to use prefabricated beams as ledgers with additional suspension ropes.

Ledgers

Ledgers hung from the wire ropes either by being passed through soft eye splices in the ends of the wire rope, or with a round turn and two half hitches with the ends seized with a minimum of two bulldog grips.

A right-angle coupler should be fixed on each side of the eye or hitch to keep the wire rope in place.

Joints in ledgers should be made with sleeve couplers and a short tube spliced on with parallel couplers or two universal couplers or, if this is not practical, longer ledgers may be used provided each ledger is underslung from at least two transoms.

Transoms should be spaced on the ledgers at intervals to suit the boards in use.

Erecting

The contractor must first check the anchorage points to ensure that they are strong enough to carry the weight of the slung scaffold and the load it will carry with due regard to any shock loadings.

The scaffolder should check all his materials for serviceability before use, especially the wire ropes. The ropes should show no sign of kinking, and there should be no broken wires in the lay.

Sequence of erection

Assume the size of the job to be 4.2 m × 4.2 m and 1.8 m drop using wire ropes.

1. Flush ends of 2 tubes and mark off 300 mm from flushed ends, measure and mark for 2.1 m spacings. Secure 3 wire ropes to one of the measured tubes. Secure wire ropes to

prevent slipping. Fix wire ropes to suspension points to form a trapeze, level and fix (Fig. 9.1).

2. Secure 3 wire ropes to second measured tube and secure to prevent slipping. Fix wire ropes to suspension points to form second trapeze, secure wire rope temporarily. Place transoms across the two trapezes or ledgers and level ledgers. Secure all wire rope (Figs. 9.2).

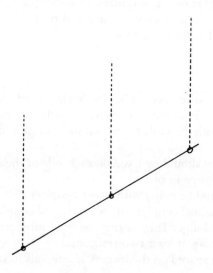

Fig. 9.1

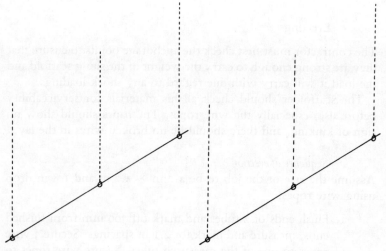

Fig. 9.2

3. Place and secure transoms. Place boards for working on (Fig. 9.3).
4. Undersling an intermediate ledger, and secure ledger to suspension points with 3 wire ropes (Fig. 9.4).
5. Deck out the platform, erect puncheons to secure guardrails and tie scaffold to prevent undue movement (Fig. 9.5).

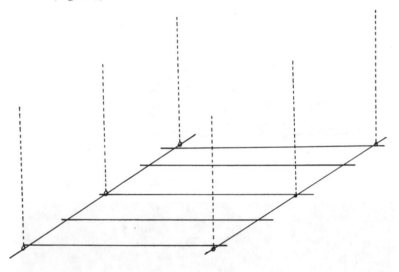

Fig. 9.3

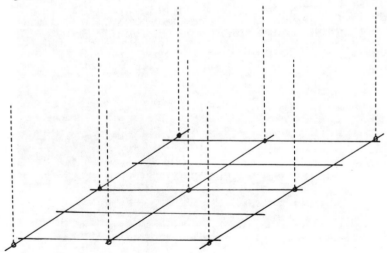

Fig. 9.4

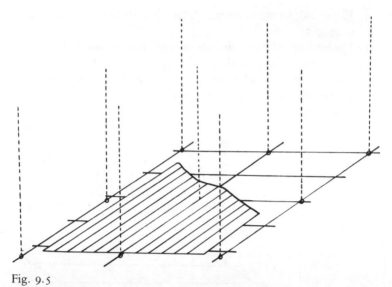

Fig. 9.5

Figs 9.1–9.5 Sequence of erection for a slung scaffold

Fig. 9.6 Slung scaffold

Fig. 9.7 A round turn, two half hitches and three bulldog grips

Safety checklist

1. Supporting members of adequate strength.
2. Wire ropes secured correctly.
3. Correct size and strength of wire rope and bulldogs the right way round.
4. Suspension points correctly spaced and fixed.
5. Suspension wires all taut before and after loading.
6. Ledgers and transoms correctly spaced and fixed.
7. Boards evenly spaced and properly supported.
8. Scaffolds tied to prevent movement.
9. Guardrails and toe-boards properly fixed.
10. No overloading of scaffold.

Fig. 9.8 Securing wire rope to ledgers

Checklist (slung scaffold)

1. Is there a strong and secure anchorage for the suspended platform.
2. Have the roof structure and its cross girders been inspected to ensure they are capable of bearing the weight of the platform and the men working on it?
3. Are the wire ropes strong enough to support the platform?
4. Are the wire ropes protected from any sharp edges by suitable packing material?
5. Are the wire ropes taut?
6. Have the wire ropes been properly secured with a round turn and two half hitches?

7. Have the ends of the wire rope been properly secured with at least two bulldog grips?
8. Are the wire ropes evenly spaced?
9. Are the ledgers and transoms fixed with right-angle couplers?
10. Is the working platform closely boarded and properly supported?
11. Has the working platform been secured to prevent movement?
12. Are guardrails and toe-boards provided?

10

Mobile towers

Description and uses

A mobile tower is a scaffold mounted on wheels or castors. It usually consists of four standards, and is square in construction.

It should not have more than one working platform and guardrails and toe-boards must be provided. Access may be gained to the working platform via a ladder or stairway positioned from either the inside or the outside of the structure.

A mobile scaffold tower must only be used for lightweight work, e.g. painters, plumbers, etc.

Construction

The height of the lifts should not exceed 2.7 m or be greater than the minimum spacing between standards. The base lift should be fixed as near to the castors as possible.

Bracing should be fixed on all four sides with the bracing at an angle of approximately 45° to the horizontal.

Ledgers and transoms must be fixed to the standards with right-angle couplers, the bracings to the ledgers and transoms by right-angle couplers or to the standards with swivel couplers, and plan braces fixed to standards by right-angle couplers. Standards should be joined with sleeve or spigot couplers with the joints staggered.

Operation

No persons, equipment or materials should be on the working platform or elsewhere on the structure when it is being moved.

The wheel brakes must be secured when the tower is in use.

Foundations

To avoid instability a mobile tower should only be moved on a firm and level surface.

If the ground is uneven or soft it is necessary to lay a temporary foundation or track to make it easier to erect and move the tower.

The track must be soundly constructed and anchored so that the bearing capacity of the ground immediately below the track is not exceeded at any point due to the imposed loading from the scaffold. The track must be anchored to the ground to prevent movement.

Base to height ratios

(a) Internal. The maximum height of an internal mobile tower should not exceed three and a half times the smallest base dimension.

No tower should be built with a base dimension of less than 1.2 m.

(b) External. The height of external mobile towers should not exceed three times the smallest base dimension.

No tower should be built with a base dimension of less than 1.2 m.

Rigidity

Towers must be maintained rigid in plan by the use of plan bracing placed at the base lift, every alternate lift, and under the working platform.

Castors

Castors or wheels should have the safe working load stamped on them and care should be taken to ensure that the weight of the tower and the distributed load does not exceed this.

Castor wheels must be fixed to the corners of the tower in such a way that they cannot fall out when the tower is moved and a wheel is out of contact with the ground.

Castors are usually about 125 mm in diameter and must not be any less. Each castor must be fitted with an effective wheel brake which cannot be released accidentally (Fig. 10.1).

Fig. 10.1 Castor wheel (mobile tower)

Loading

The tower should be constructed to support, in addition to its own height and the weight of the scaffold boards, a distributed load of 150 kg/m^2 over the working platform.

Ladder access

When the means of access to the working platform is fixed to the outside of the tower, care should be taken regarding the effect this may have on the stability of the tower.

Sequence of erection (using GKN fittings)

To erect a mobile tower with dimensions 2.4 m × 2.4 m and 3.66 m high, for example, the following sequence may be used:

1. Select 4 tubes, 2 of which will be used as ledgers, and 2 of which will be used as transoms, and lay them down with the ends in line. Measure from one end of each tube 150 mm and mark all 4 tubes. Measure 2.4 m from these marks and mark again.

2. Fix 2 right-angle couplers to each ledger and transom on centre of marks (Fig. 10.2).
3. Erect first standard and fix 1 ledger and 1 transom to standard. Ledger must be fixed to inside of standard, and transom fixed above ledger also on inside of standard, approximately 150 mm clear of ground to permit location of castor wheels (Fig. 10.3).
4. Erect second standard to transom, level transom and fix (Fig. 10.4).
5. Erect second ledger to standard underneath transom and fix (Fig. 10.5).
6. Erect third standard to ledger, level ledger and fix to standard (Fig. 10.6).

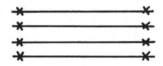

Fig. 10.2

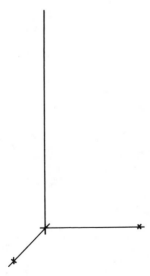

Fig. 10.3

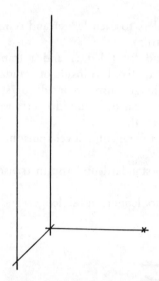

Fig. 10.4

Fig. 10.5

Fig. 10.6

7. Fix second transom to standard, above ledger (Fig. 10.7).
8. Fix fourth standard to ledger and transom, level ledger and fix ledger and transom (Fig. 10.8).
9. Measure 1.8 m from ledgers and mark all 4 standards. Fix 2 right-angle couplers to each standard, one to receive ledger, and one to receive transom. Place ledgers and transoms in couplers (Fig. 10.9).
10. Place braces in right-angle couplers fixed to ledgers and transoms, or swivel couplers fixed to standards. Braces should be fixed diagonally opposed on opposite sides. This method of bracing is sometimes referred to as 'chasing the brace'.

    Secure right-angle or swivel coupler at bottom of brace, plumb adjacent standard and secure top of brace to standard. Repeat on all 4 sides. Secure the 2 ledgers and 2 transoms. Plumb standards (Fig. 10.10).
11. Fix a plan brace from standard to standard with right-angle couplers. Fix castor wheels, secure to standards, secure brakes on castor wheels (Fig. 10.11).
12. Fix an intermediate transom to ledgers with 2 putlog couplers. Sheet out platform (Fig. 10.12).

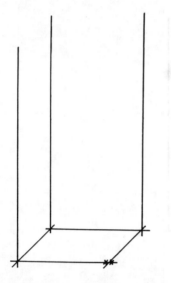

Fig. 10.7

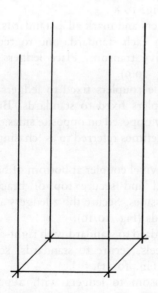

Fig. 10.8

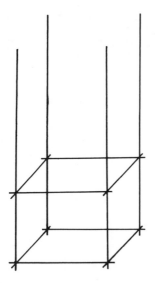

Fig. 10.9

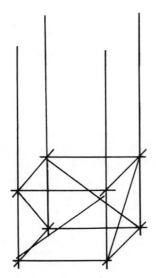

Fig. 10.10

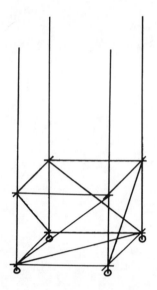

Fig. 10.11

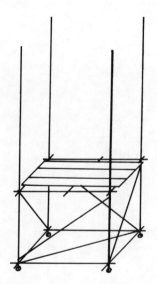

Fig. 10.12

106

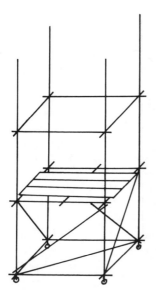

Fig. 10.13

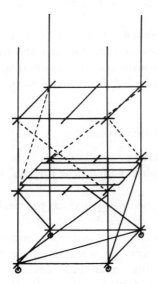

Fig. 10.14

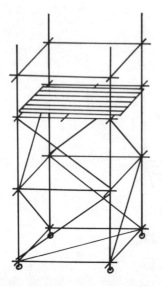

Fig. 10.15

Figs. 10.2–10.15 Sequence of erection for a mobile tower

13. Measure 1.8 m from ledgers, fix 2 right-angle couplers to standards, place ledgers and transoms in couplers (Fig. 10.13).
14. Place braces in right-angle or swivel couplers, plumb standards and secure braces, secure ledgers and transoms, brace all 4 sides. Fix an intermediate transom (Fig. 10.14).
15. Measure and fix guardrails, sheet out platform and fix toe-boards (Fig. 10.15).

Safety checklist (mobile towers)

1. Are castor wheels secured to standards?
2. Are castor wheels supplied with efficient brakes?
3. Is ground level?
4. Is foot-tie lift as close to castor wheels as practical?
5. Are joints in standards staggered?
6. Are ledgers and transoms secured to standards with right-angle couplers (except on working platform)?
7. Are braces fixed to all lifts, on all four sides?
8. Are plan braces fixed correctly?
9. Is height of scaffold within base × height ratio?

Fig. 10.16 Mobile tower

10. Is working platform close-boarded and properly supported?
11. Are guardrails and toe-boards properly fixed?
12. Is ladder sound and properly secured?
13. Does ladder extend above working platform by at least 1.06 m?

Fig. 10.17 Static tower

14. Is scaffold overloaded?
15. Does lift height exceed minimum spacing between standards?

11

Hoists

'Hoist operator caused gate to descend on head of labourer' (head injuries). 'Hoist operator inadvertently pulled rope as labourer was stepping off hoist, causing hoist to drop 1 m and man to fall over barrow on hoist platform' (fractured bone in wrist). 'Labourer bending down placing scaffold frames ready for loading with rear part of body inside hoist area, was struck by the descending hoist' (deep bruising to side of chest and leg). These are three typical accidents with hoists. Men have been known to stick their heads into the hoist tower to see when the platform was due to descend, and were quite surprised to find that they were struck by the hoist platform. **Inadequate or non-existent signalling procedures, the hoist striking projecting tubes, or the load not being properly secured, can all contribute to the cause of accidents**.

It must be realized that under the Construction Regulations a mobile or static cantilever builders' materials hoist, whether worked by mechanical means or otherwise, is a 'lifting machine' for raising or lowering materials by means of a moving platform 'restricted by guides'.

A hoist may have as its *power source* either a diesel/petrol engine or an electric motor with a friction winch or an electric geared winch.

The *location* of the hoist is most important and the siting of the hoist requires careful planning; it should be sited within easy distance of the stacking area, and so that access is available to all levels of the building. A firm foundation is necessary and the ground should be well consolidated with no voids. It may be necessary to get the general foreman to lay a lean concrete mix over the area, and perhaps railway sleepers or RSJs to act as soleboards.

Erection procedure

In addition to a manufacturer's or supplier's Certificate of Test a manufacturer or supplier will issue an Erection Procedure and this erection procedure must be strictly adhered to.

1. Before erecting the mast, check that the mast sections are of the same gauge and that the bolt holes are not damaged or distorted. Make sure the bolt holes marry up.
2. Check the maximum height to which the hoist may be erected, (the manufacturer will make this recommendation). It is usually about 6 m for mobile hoists but may vary slightly.

   If the mast is one which is tied to a scaffold tower or building the maximum recommended height is usually about 45 m.
3. Position the platform at right angles to the building and, if levelling jacks are provided, level hoist and tension stay wires.

Hoist operation

As an example, one type of hoist uses a rope fixed to the brake, the rope being passed over a pulley at the top of the mast and back down the mast to the brake.

When the brake handle is raised by giving a slight pull on the rope this centres the drum between the brake and the driving shaft, allowing the drum to rotate freely and the hoist platform to descend. A strong pull on the rope forces the drum on to the driving shaft which is turning continuously. This causes the drum to rotate in an anti-clockwise direction thus heaving in on the wire and thereby raising the hoist platform. If the rope is let go, the drum is held hard to the brake shoe by counterweights. This stops the drum from revolving and holds the platform. In case the hoist wire between the winch and the platform should break, the platform is fitted with a locking device which will automatically engage once the tension in the wire is released.

Hoistway

This is the framework of tubular scaffolding designed to support the track or mast guides of the hoist.

1 Over-run device
2 2 m high landing gates
3 Hoist enclosure 2 m minimum height
4 Hoist mast tied into building
5 Hoist arrestor device
6 Hoist operated from one position only, giving
  driver unobstructed view
7 Dead man handle
8 S.W.L. marked on hoist platform

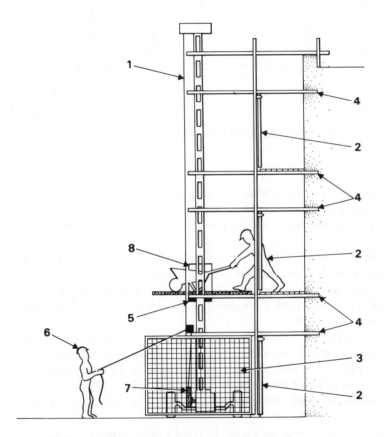

Fig. 11.1   Sketch of goods hoist illustrating the requirements of
           Regulations 42 and 44 of the Construction (Lifting Operations)
           Regulations 1961. The wire mesh surround to the hoistway has
           been omitted in order to allow detail to be shown

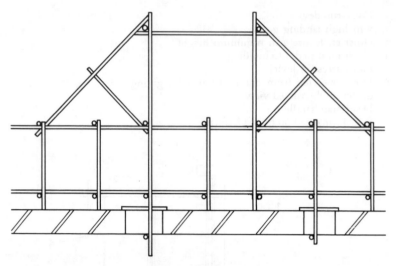

Fig. 11.2 Plan of bracing and through ties for hoist

The hoist mast is usually box-shaped and is constructed of mild steel angle.

The foundations must be levelled and compacted and if the bearing capability of the ground is in doubt, then concrete, hardcore or other suitable material must be provided. All standards must be pitched on steel baseplates and soleboards.

If the lifts have to be erected to suit specific levels, these levels must be determined beforehand to avoid time-consuming and expensive alterations.

The internal dimensions of the tower will depend on the size of the hoist platform and the loading area. Standards at the corners of the tower may be either single standards or in twos, threes or fours, according to the loads required or specified. All standards must be plumbed and joints staggered and spaced at no more than 2.1 m apart.

Ledgers and transoms should be fixed so as to project beyond the standards by about 300 mm on each side to allow for fixing the braces. The braces should be secured to the ledgers and transoms with right-angle couplers. Bracing must be fixed on all four sides of the tower (except across access points) with right-angle couplers. Plan bracing should be fixed from the corner standards back to the building at every alternate lift and be secured to the building with two-way ties. This will reduce vibration and stabilize the tower. Plan bracing should be kept as short as reasonably practical and in any case not more than 4.25 m in length.

Two-way ties must be fixed to secure the hoist tower to the building at every floor level. Any hoist tower erected higher than 64 m or which cannot be tied to the building must be specially designed.

No hoist tower must ordinarily extend more than 6 m above the highest tie, tube or bracing.

Plan braces must be fitted at the top of the hoist tower, clear of the movement of the hoist.

Platform

It is essential that the platform is well constructed with no splits or damage and it is good practice when erecting the supporting tower to allow 50 to 75 mm clearance all the way around for the platform.

The platform rollers should be inspected to make sure that they are the right size for the mast rails, that they are not cracked and have no flat spots. Rollers must be free from grease. It is surprising how many people think that all moving parts should be greased and on that basis grease the rollers and the mast.

There are serrated cams on the platform. They must be cleaned regularly, brushed with a stiff brush to make sure they are not clogged with dirt or mud, and checked to see that the spring is properly tensioned. This will be very important should the platform accidentally free-fall.

The hoist rope should be inspected for wear, size and length. It should be fitted with a thimble for securing the rope to the mast head anchorage. There should be three bulldog grips used, and they should be secured correctly, i.e. properly spaced (six rope diameters apart) with the nuts of the bulldog grip on the 'live' end of the ropes.

In order to help the hoist operator to judge the landing positions, short lengths of rubber hose may be fixed to the edge of the platform, so that when the operator sees the short length of rubber hose come into contact with the landing place, he will know that the platform is level with the landing place.

Hoist motor

The manufacturer's recommendations regarding the distance

between the hoist motor and the tower should be adhered to, and it is most important that the hoist motor is positively anchored to prevent the motor being dragged to the tower when the hoist is being operated. It is also important that the hoist motor is in line with the tower. If it is not, then this may cause a side pull on the hoist rope. There should be at least two full round turns left on the drum when the platform is at its lowest level.

Control gear

An over-run device known as a 'bob-weight' or 'egg-stop' is usually fitted to prevent the hoist platform from over-running the hoist mast. Where an egg-stop is fitted, the ring for the egg-stop must not be so big as to allow the 'egg' to pass through. The egg should be firmly secured to the hoist rope at approximately 1 m from the top of the mast head and the hoist rope not tensioned too much or left too slack.
Leaving the rope too slack prevents the device from operating.

Guards

Gates must be fitted to all landings and warning notices KEEP THIS GATE CLOSED and RIDING PROHIBITED fixed to all gates. All gates must, in any case, be easy to open and close and the bottom of the gate must be prevented from swinging inwards. A further notice ONLY AUTHORISED PERSONS TO OPERATE HOIST should also be displayed.
    It is also necessary to enclose the base or first lift of the tower with some substantial fencing for protection, such as corrugated iron sheets, and the rest of the tower above this height should be enclosed with wire netting.

Operation

Some consideration must be given to the facilities for the operator, e.g. if the hoist is to be operated from the top landing, a trussed-out platform is required which must be properly constructed together with guardrails and toe-boards at both sides and ends. There must be guardrails fixed between 914 mm and 1.15 m and toe-boards a minimum of 155 mm high, leaving a gap of not more than 750 mm between the top of the toe-board and the underside of the guardrail.

All pulleys, gears, chains, belts, etc, must be fitted with suitable guards.

Protection from materials falling and inclement weather must be provided for the hoist motor, and all-weather electrical equipment, switches and wiring should be used and a lock-off isolating switch provided.

All pneumatic tyres of mobile hoists must be correctly inflated. Where work is to be carried out during hours of darkness or in poor light, temporary artificial lighting must be sufficient and suitable.

Engine exhausts must be fitted with effective silencers and electric motors screened to reduce noise to an acceptable level.

Where a hoist is located within a building, the exhaust pipe of the internal combustion engine should be extended to the outer atmosphere.

Should it be necessary to work from the platform during the erection of the hoist, the platform must be properly supported on good strong cross-bearers, not scaffold boards!

Communication is often a problem between the ground and the upper landing place, especially on high-rise buildings, so some consideration should be given to the purchasing or hiring of a two-way communication system.

If it is necessary at a later date to extend the mast, the new rope must be long enough to do the job, plus two round turns on the drum.

Operator

All operators must be over the age of eighteen, unless they are under direct supervision. They should be medically fit, have good eyesight and hearing and be able to judge distances. It is essential that the operator receives proper training in his duties. This may be done by an experienced operator on site. The operator should make sure the trainee operator has a copy of the manufacturer's instruction regarding the operational procedure, knows what to do in an emergency and is able to carry out routine maintenance. It should also be made clear to the trainee operator and everyone else on site that the **operator is solely responsible for the safe operation of the hoist**. A signboard to this effect should be displayed on or near the hoist.

The trainee operator should be instructed to carry out, at the beginning of each day, a check on the controls of the hoist according to manufacturer's recommendations.

In wet weather the controls may be tested by making a dummy run with the platform.

Any defects must be reported without delay to the supervisor.

When it comes to operating the hoist the trainee operator should also be instructed as to the maximum loading permissible on the platform and, although the SWL must be displayed, the trainee operator may not be sure of the weight of the materials he will be asked to raise or lower.

He must be told that all materials must be properly secured and well clear of the edges of the platform. Any barrows that are being used must be positioned with their handles facing the exit, so that anyone removing the barrows from the platform will be forced to do so without walking on to the platform. It would be quite easy for the hoist operator to mistake a body movement for a signal to lower the platform, with the result that the barrow man might fall.

Before operating the hoist, the operator must check that all gates are closed and the hoistway is clear of obstructions.

It is important that, when a hoist is raised or lowered, it is done smoothly and controlled by using the winch brake with the hoist rope under tension. This will prevent any snarling of the rope. The platform must not be allowed to free-fall, but must be controlled to land gently on the tyre buffer to prevent damage to the platform.

The operator must be told that he must never leave the controls when the platform is suspended and, before going off duty, he must lower the platform to the ground, switch off the engine and if possible isolate to prevent unauthorised use.

Safety checklist (hoist towers)

1. Are ground conditions suitably firm to carry superimposed load of scaffold and loads on hoist tower?
2. Are soleboards long enough, wide enough and thick enough to distribute loads properly?
3. Are all standards pitched on steel baseplates?
4. Are all standards plumb?
5. Are all joints in standards staggered?
6. Are joints in standards positioned as close to ledgers as possible?

7. Are ledgers and transoms level and fixed to standards with right-angle couplers?
8. Are lift heights in excess of 2.13 m?
9. Are plan braces fixed at every floor level?
10. Is diagonal bracing fixed to ledgers and transoms with right-angle couplers?
11. Is diagonal bracing fixed to all four sides, from base to full height of scaffold?
12. Does hoist extend more than 9.75 m above highest bracing?
13. Is hoist tower tied to both main scaffold and building at every floor level?
14. Are there sufficient and suitable ties?
15. Has hoist tower been completely enclosed on all four sides with wire mesh (except for landing places)?
16. Have gates been fitted at landing levels?
17. Is there a proper and suitable hoist driver's platform?
18. Has hoist driver's platform been provided with suitable guardrails and toe-boards?

12

System scaffolds

There is available a large range of alternatives to the tube and coupler scaffold. Each in its own way may provide the best solution to a specific problem.

Some of the alternative methods available are:

(a) *lorry-mounted hydraulic arm platform* – suitable for one-off access, although the height is limited.

(b) *telescopic mast* – a one-man access platform with limited height.

(c) *modular tubular metal systems* – either in steel or aluminium alloy with components linked together, usually without using normal couplers, to provide a structure that replaces a normal tube and coupler scaffold.

It has been said that the United Kingdom uses more tubes and fittings than the rest of the world put together, whilst in the USA. and Europe they favour the modular systems.

When comparing 'system' scaffolds with tube and fittings scaffolds, the following factors should be considered. Considerable erection time may be saved thus saving labour costs. Components are simple and easily located and, provided the base lift is plumbed and levelled and squared, the scaffold needs no further setting out or plumbing because of its modular nature.

Vertical joints between frames are fixed and usually of the spigot type with means of resisting tension with:

(a) self-retaining pins;

(b) shaped, twist grip locking devices; or

(c) spring clips.

There are dozens of different frame shapes available using different tube sizes, but the majority use the normal 48 mm tube for the main vertical members. They fall into three groups:

(a) access scaffold frames;
(b) shoring scaffold frames;
(c) access and/or shoring frames.

The 'H' frame system

Mainly for access purposes, this system has only one basic unit – the 'H' frame. The 'H' frame is a welded steel frame roughly 'H' shaped. It has brackets attached to the vertical and horizontal members and these receive the normal tubes which connect one frame to the next. This system has two main advantages:

(a) The centres of the frames are adjustable in the longitudinal direction.
(b) The welded joint between standard and transom reduces the need for ledger bracing.

However, this type of open portal frame should be adequately and properly braced and tied if used for very high or very heavily loaded scaffolds.

Tube component modular system

This system does away with loose couplers and diagonal bracing.

(a) Standards are fitted with tension spigots and welded-on attachments at set intervals which receive:
(b) ledgers and transoms of fixed length fitted at the ends with attachments which are locked on to those on the standards. This also reduces the number of diagonal braces.

Whilst this system allows for the lift heights to be varied, it does limit the bay size, owing to the limited range of horizontal members.

The connections are fixed in several ways, one of which is a locking screw; another is a bolt; or the connection may be made with hammered wedge action.

The system may be used for a support scaffold or an access scaffold.

Diagonal bracing must be used for both access and support scaffolds carried out in this system.

Frame and brace system

This system consists of two basic components:

(a) welded tubular frames which fit via spigots one on to another;

(b) the cross brace or scissor brace, which consists of two lengths of light angle or tube, pivoted at mid-length. There is a hole at each free end to fit over a stud with a quick locking device projecting from the legs of the frames. This module results in a very stiff scaffold, owing to the large amount of bracing. It is also easy to adjust any settlement of the scaffold standards by use of the adjustable bases.

Fig. 12.1 'Allround' scaffold system

The rectangular frame system

This consists of only one basic unit – a rectangular frame, several of which may be fitted together to provide a square or rectangular tower. The main structure is composed of a number of rectangular frames placed one above the other, alternately at 90° on plan, and plan braces or plan corner stiffeners must also be used.

The system may be used to provide access or falsework towers or several towers may be linked to form a birdcage scaffold.

The triangular frame system

This system is a very efficient falsework scaffold. The key to its strength is the fact that, being constructed of triangular frames, the

Fig. 12.2 'RMD' scaffold system

Fig. 12.3 'Cuplok' scaffold system

scaffold is very well braced diagonally and has little chance of failure from insufficient lateral restraint.

A large variety of plan arrangements for triangular and/or square towers may be used, giving a variation in deck support.

As with all tower systems, groups of towers should be linked at intervals of not more than four lifts in height and diagonally braced.

13

The Scaffolders Record Scheme

In March 1978 the Construction Industry Training Board formally approved the Scaffolders Record Scheme with its arrangements for training and certification of operatives.

The administrative details of the scheme were then published for the industry stating that there would be a trial period during 1979 (1 January 1979 to 31 December 1983). It is interesting to note that 21000 operatives were registered during this period, sufficient justification for the decision to go ahead with the scheme.

The scheme has four main benefits:

1. by ensuring a certain standard of training and experience of scaffolders in the industry, it should lead to improved standards of safety;
2. it will provide employers with the benefit of knowing that a man who has a particular scaffolder's card has received training and/or has experience to a prescribed level;
3. it will provide the scaffolder with status relevant to his importance within the construction industry;
4. it will ensure that the training provided by the Construction Industry Training Board will reach the standard required.

The scope of the scheme

The Construction Industry Scaffolders Record Scheme applies to all operatives who are employed under the agreements of the National Joint Council for the Building Industry or the Civil Engineering Construction Conciliation Board who are at any time required to erect, substantially alter or dismantle scaffolding, with certain exceptions noted below. The Scheme came into operation from Monday 1st January 1979.

The working rules of the NJCBI and CECCB provide that an operative who has not attained prescribed levels of training and experience in scaffolding work of a given kind must not be employed on such work unless:

1. under adequate supervision; or
2. working with an operative who has the required training and experience; or
3. erecting, altering or dismantling simple access scaffolding with a working platform no higher than 5 m.

The required training in this context means the satisfactory completion of a course approved by the NJCBI and/or CECCB in one of the following:

1. Basic scaffolding operations (see list in Appendix A).
2. Advanced scaffolding operations (see list in Appendix A).

That is, training courses in Basic and Advanced scaffolding, with the content approved by the NJCBI and/or CECCB, to be provided by the Constrruction Industry Training Board or by other training centres with CITB approval.

The important part of the scheme is the training standards achieved and maintained by the individual scaffolder and, although for the start of the scheme this was based on the recommendations of the employers, for the future this will depend upon the scaffolder attending formal training courses run by the National Training Centres or bodies approved by the CITB and by gaining, through experience, training in those aspects not covered in the formal training.

The training element

There are three operative courses.

The trainee scaffolder attends two two-week courses: Basic Part One and Basic Part Two – with a minimum of six months related experience between the two courses, and between twelve and eighteen months overall experience before receiving a Basic Scaffolder Training Record Card.

The basic scaffolder must undertake a further two weeks training coupled with a minimum of one year's experience before qualifying as an advanced scaffolder.

The CITB has prepared a completely new two-week course for the advanced scaffolder and although a trainee will now undertake six weeks of training before qualifying as an advanced scaffolder, not every item listed in the approved list of scaffolding operations is covered in the off-the-job training periods, i.e. certain operations, particularly those of a very specialist nature, have to be learnt on the job.

The contents of the CITB scaffolding courses are now all set out in objective output terms, clear statements of what the trainee will know and be able to do at the end of his training:

Basic Part One

At the end of a scaffolder's Basic Part One course the trainee should be able to:

(a) erect and dismantle correctly an independent scaffold with return and fan;
(b) erect and dismantle correctly a mobile tower;
(c) inspect, erect and lash pole ladders;
(d) inspect, fix and reeve gin wheels;
(e) erect and dismantle correctly a putlog scaffold with return;
(f) erect and dismantle correctly a birdcage scaffold.

Fig. 13.1 Independent scaffold

Fig. 13.2 Dead or needle shore scaffold

In addition to his skill, the trainee should know:
  (a) the types and uses of scaffold fittings, tubes and boards in common use;
  (b) building construction and its processes relevant to the erection of scaffolds;
  (c) Construction and Statutory Regulations relevant to the erection and dismantling of basic scaffolds.

Fig. 13.3 Gantry

Basic Part Two

Scaffolder Basic Part Two is for a trainee scaffolder who has
successfully completed the Basic Part One course and has had at
least six months planned on-site experience.

Fig. 13.4 Falsework

At the end of this course a trainee should be able to:

(a) erect and dismantle correctly a heavy duty independent scaffold with bridging;
(b) erect and dismantle correctly a truss-out scaffold;
(c) erect and dismantle correctly a builder's hoist;
(d) fix netting to hoist towers and various types of sheeting to scaffolds;
(e) erect and dismantle correctly a cantilever drop scaffold;
(f) erect and dismantle correctly a stack scaffold;
(g) erect and dismantle correctly a tubular slung scaffold.

In addition to his skill the trainee should know:

(a) the types and uses of scaffold fittings, tubes and boards in common use;
(b) Construction and Statutory Regulations relevant to the erection and dismantling of basic and complex and designed scaffolds.

130

Fig. 13.5 Independent with bridge

Advanced Scaffolding

The Advanced Scaffolder course is for a basic scaffolder who has had a minimum of six months planned on-site experience as a basic scaffolder. At the end of this course the trainee should be able to:

    (a) erect scaffolds in accordance with drawings and specifications;

131

Fig. 13.6 Independent with bridge and buttress

    (b) erect and dismantle correctly a dead shore;

    (c) erect and dismantle correctly a raking shore;

    (d) rig and dismantle correctly rope or winch-operated boatswains chairs and cradles;

    (e) fix and dismantle correctly guys and anchorages for scaffolds;

    (f) erect and dismantle correctly lifting structures complete with runway beams;

    (g) inspect scaffolds for faults.

At the end of this course the trainee should know:

    (a) the types and uses of scaffold fittings, tubes and boards in common use;

    (b) Construction and Statutory Regulations relevant to the erection and dismantling of basic, complex and designed structures;

    (c) how to read and interpret correctly scaffold drawings and specifications;

    (d) how to inspect scaffolds for faults.

The trainee's progress is measured during each course. The practical training is built up of a number of progressive exercises covering each of the practical objectives and the trainee is assessed on each exercise.

In addition to his practical assessment, a trainee is required to answer a written question paper.

14

The Construction (Working Places) Regulations 1966 – applying to scaffolds

Scaffolding used in building and civil engineering is governed by The Construction (Working Places) Regulations 1966. This chapter lists the Regulations and indicates their main practical effect, but should not be taken as a definitive statement. Those responsible for the supervision and management of scaffolding operations should possess a copy of the statutory instrument itself (1966 No. 94) available from Her Majesty's Stationery Office. All of the procedures described in this book have been developed to comply with the Regulations. Amplification and explanation will be found in the text of preceding chapters. Readers are also referred to *Site Safety* by J.C. Laney, published in this Site Practice series. The Regulations have been metricated by the Construction (Metrication) Regulations 1984.

Part I: Application and interpretation

Regulation 1: Citation, commencement and revocation
Sets the current Regulations into context with earlier instruments.

Regulation 2: Application of Regulations
The Regulations apply to all building work and most engineering construction.

Regulation 3: Obligations under Regulations
It is the duty of all employers and employees to comply with the Regulations and to report any defect in plant or equipment which may be discovered.

Regulation 4: 'Interpretation'
The Regulations assign specific meaning to terms used in the
Regulations.

- 'ladder' does not include a folding step-ladder;
- 'ladder scaffold' is a working platform supported directly
  on a ladder by crutches or brackets;
- 'lifting appliances' are any equipment for raising or
  lowering, also excavators, piling frames, draglines and
  aerial ropeways or runways;
- 'lifting gear' is ropes, chains, shackles and the like;
- 'plant or equipment' means any plant, equipment, gear,
  machinery apparatus or appliance;
- 'scaffold' means any temporarily provided structures on or
  from which persons perform work;
- 'sloping roof' is any roof pitched at an angle greater than
  10 degrees from the horizontal;
- 'slung scaffold' means a scaffold suspended by means of
  lifting gear;
- 'suspended scaffold' means a scaffold similarly hung but
  which is capable of being raised or lowered;
- 'trestle scaffold' is a platform supported by self-supporting
  step-ladders, split heads or any such moveable
  contrivances;
- 'working platform' includes a working stage.

Part II: Exemptions

Regulation 5: Certificates of exemption
The Chief Inspector may by certificate in writing exempt from all
or any of the Regulations if he is satisfied that, in a particular case,
the Regulation is either unnecessary or not reasonably practical.

Part III: Safety of Working Places and Access and Egress

Regulation 6: General
Every place where a person works shall be kept safe and there must
be safe access to and egress from every place.

Regulation 7: Provision of scaffolds, etc
Sufficient and suitable scaffold, ladders or other means of support

shall be provided where work cannot be done from the ground or part of a building.

Regulation 8: Supervision of work and inspection of material
All material for any scaffold must be inspected by a competent person before use and no scaffold may be erected, dismantled or altered except under the immediate supervision of a competent person and as far as possible by competent, experienced workmen.

Regulation 9: Construction and material
(1) and (2)   All scaffolding shall be well constructed of suitable, sound, adequately strong and sufficient material.

(3) and (4)   Timber shall be of suitable quality, in good condition and stripped of its bark. It may not be painted or otherwise treated so that defects would be concealed.

(5)   Metal parts shall be suitable, in good condition and free from corrosion or other patent defects likely materially to affect their strength.

Regulation 10: Defective material
(1) and (2)   No defective material, rope or bond shall be used.

(3)   Materials and parts fit for use shall be kept in suitable conditions when not in use and apart from defective material and parts.

Regulation 11: Maintenance of scaffold
Every scaffold shall be properly maintained and every part kept fixed, secured or placed in position so as to prevent accidental displacement.

Regulation 12: Partly erected or dismantled scaffold
Except where partly erected or dismantled scaffold complies with the Regulations any parts which are accessible shall be blocked off or have a prominent warning notice fitted. Otherwise the scaffold must be completed until it is safe or dismantled until it is inaccessible.

Regulation 13: Standards or uprights, ledgers or putlogs
(1) and (2)   Standards shall be vertical or inclined slightly towards the building, sufficiently close together and placed on an adequate base to prevent slipping or sinking.
(3)   Ledgers shall be horizontal and securely fixed to the standards.

136

(4) and (5) Putlogs shall be securely fastened to ledgers or standards or their movement otherwise prevented and where supported by one end on a wall there shall be sufficient bearing. The distance between putlogs shall have regard to the expected loading and should not exceed 1 m for 32 mm boards, 1.50 m for 38 mm or 2.60 m for 50 mm boards.

Regulation 14: Ladders used in scaffolds
(1) and (2)   Ladder scaffold shall only be used for light work, must be adequately strong, evenly supported on or suspended from each stile and secured against slipping.

Regulation 15: Stability of scaffolds
(1) and (2)   Every scaffold shall be secure and stable in itself or sufficiently braced and tied to the building or other supporting structures. Such supporting structure shall be sound and sufficient to prevent collapse and ensure stability and, if necessary shall itself be strutted or braced.
(3)   Mobile scaffolds shall be stable and, if necessary, broader at the base, on a firm, even and level surface; secured to prevent movement when in use; and move only by pushing or pulling at or near the base.
(4)   Except for platforms no more than 600 mm high scaffolds may not be supported on loose bricks or other building materials.

Regulation 16: Slung scaffold
(1)   All chains, ropes, lifting gear and metal tubes shall be suitable and adequately strong; properly secured to safe anchorage points and to the scaffold member; placed to ensure stability; vertical; and kept taut.
(2)   No rope other than wire rope may be used.
(3)   Chains and ropes must be prevented from coming into contact with edges where this would cause danger.
(4)   Slung scaffolds must be prevented from swaying unduly whilst in use.

Regulation 17: Cantilever, jib, figure and bracket scaffolds
(1)   Cantilever and jib scaffolds must be adequately supported, fixed and anchored; have outriggers of adequate length and strength; and be strutted and braced to ensure rigidity and stability.
(2)   Dogs, spikes or similar fixings may not be used to support figure or bracket scaffold if they are liable to pull out.

Regulation 18: Support for scaffolds, etc

No part of a building including gutters, may be used to support scaffolding, ladders or the like unless it and any fixing is strong enough to provide adequate support.

Regulation 19: Suspended scaffolds (not power operated)

(1) and (2) The Regulation applies to both temporary and permanent suspended scaffolds.

(3) The chains or ropes and winches shall be suitable, adequate and suspended from safe anchorages.

(4) Winches shall be fitted with automatic (fail-safe) brakes and protected from the weather, dust etc.

(5) and (6) Outriggers shall be adequately long and strong, properly spaced and supported, provided with stops at their ends and installed horizontally. Where counterweights are used with outriggers the counterweights shall be securely attached to the outriggers and shall be not less in weight than three times the weight which would counter-balance the total weight suspended from the outrigger, including runway, joist, suspended scaffold, persons and other load. (See (15) for exceptions).

(7) Slung working platforms must be adequately clear of the structure.

(8) Runways, joists or tracks shall be entirely adequate and sound, securely fixed and provided with adequate stops.

(9) Ropes or chains must be adequately secured top and bottom and kept taut.

(10) The length of rope to be used shall be clearly marked on each winch. It shall be of sufficient length to have at least two turns remaining in the drum when lowered to the lowest position.

(11) Every part of a suspended scaffold shall be entirely adequate, properly maintained and free of corrosion and other defects. Platforms shall not be allowed to tip, tilt or sway whilst in use.

(12) Except for work of a light nature (see (15)) no rope other than wire rope may be used.

(13) Working platforms shall be close-boarded except to allow for drainage; 600 mm wide or, if used also for materials, 800 mm wide; and never used to support any higher scaffold. When used by sitting workmen the platform should be 300 mm or less from the face and may be held at the appropriate distance for working by short props or other devices (see (15) for exceptions).

(14) and (15) When fibre ropes and pulley blocks are used they should be not more than 3.20 m apart and the work should

be of such light nature that they can be used with safety. The minimum platform width shall be 430 mm.

Regulation 20: Boatswain's chairs, cages, skips, etc (Not power operated)
(1) No such equipment may be used unless it is properly constructed, maintained and free from defects; outriggers properly installed and supported; chains or ropes securely attached top and bottom; provided with means to prevent the occupant from falling out; free of obstruction; prevented from spinning or tipping; in the case of skips at least 910 mm deep; and the installation supervised by a competent person.
(2) They may not be used at all except for work of very short duration or if it is impractical reasonably to install a supported scaffold.

Regulation 21: Trestle scaffolds
(1) All trestles and materials used must be properly made of sound material, strong enough, properly maintained and free from defects.
(2) Trestle scaffold must NOT be used where a person might fall more than 4.50 m, if constructed with more than one tier where folding supports are used.
(3) If erected on a scaffold platform the width of such platform shall allow clear space for transport of materials; and the trestle must be tied down and braced to prevent displacement.

Regulation 22: Inspection of scaffolds, boatswain's chairs, etc
(1) No scaffold more than 6 feet 6 inches high may be used which has not been inspected and the inspection reported on the required form by a competent person during the preceding 7 days or since exposure to weather or other conditions which might affect its stability.
(2) and (3) There is no need to inspect a scaffold at less than 7-day intervals if it has only been added to, altered or partly dismantled. On a site of less than 6 weeks' duration the person in charge of the site may carry out the inspection if he is competent to do so. He must report the inspection within a week to his employer who must enter details on the appropriate form.

Regulation 23: Scaffolds used by workmen of different employers
Each employer must satisfy himself that the scaffold is safe.

Regulation 24: Construction of working platforms gangways and runs

(1)   All platforms more than 2 m high must be close-boarded or, where securely fixed, no more than 25 mm apart, in the case of metal decks they may have interstices not exceeding 4000 sq mm provided in neither case persons below are put at risk of objects falling through.

(2) and (3)   Platforms may not slope more than 1 vertical to $1\frac{1}{2}$ horizontal and where the slope exceeds 1 to 4, stepping laths must be fitted at suitable intervals for the full width of the run, except that gaps of 100 mm may be left open to allow barrow running.

Regulation 25: Boards and planks in working platforms, gangway and runs

(1)   Every board must be thick enough to span safely the distance between putlogs and not less than 200 mm wide or 150 mm wide if more than 50 mm thick.

(2)   A board must not project more than 4 times its thickness unless it is fixed to prevent tipping and, if fixed, the projection must not exceed the distance to which the board can safely be loaded.

(3)   Bevelled pieces must be fitted where boards overlap or otherwise present a tripping hazard except where the platform has one side around the curved surface of a structure.

(4)   Every board must rest evenly on at least three supports unless there is no risk of undue or unequal sagging.

(5)   Where work is to be carried out at the end of a wall the working platform should extend at least 600 mm beyond the wall.

Regulation 26: Width of working platforms

(1)   All working platforms more than 2 m high shall be at least 600 mm (3 boards) wide if not used for depositing materials. If materials are deposited the platform must be 800 mm (4 boards) wide with a clear passage of 430 mm (2 boards) for persons and 600 mm (3 boards) for moving materials. A platform must be 1.05 m (5 boards) wide if used to support a higher scaffold; 1.30 m (6 boards) wide if used by masons to dress stone; and 1.50 m (7 boards) wide if the masons also use trestles or higher platforms.

(2) and (3)   A platform on a ladder, scaffold or trestles; or below a roof for work to or near the roof; if used in connection with cylindrical metal structures may be 430 mm (2 boards) wide

provided it is safe and the work is of short duration; where workman sit on the edge of a platform the space shall not exceed 300 mm or where work is carried out to the face of a building the space shall be as small as practicable.

(4)   Where space limits the width of a platform, lesser widths than specified are permitted provided the platforms are as wide as is reasonably practicable.

Regulation 27: Width of gangways and runs
(1) and (2)   Where more than 2 m high, gangways and runs must be 430 mm (2 boards) wide for men only or 600 mm (3 boards) wide for men and materials except where there is insufficient space when they must be as wide as is reasonably practicable.

Regulation 28: Guard rails and toe-boards at working platforms and places
(1), (2) and (3)   All platforms more than 2 m high must be pro-vided with a sufficiently strong guard rail, securely fixed to the inside of uprights wherever possible, between 910 mm and 1.15 m above the platform and also above any raised standing place on the platform. Toe-boards or other barriers not less than 150 mm high must be fixed to prevent persons and materials fall-ing. The distance between the toe-board and the guard-rail must not exceed 765 mm.

(4) and (5)   Guard rails and toe-boards may be removed or not put in place but only for so long as may be necessary for access for persons or materials. Where guard rails may obstruct work to the face of a building they may be reduced to not less than 700 mm high. Where workers sit on the edge of a platform and secure hand holds of rope or chain are provided, guard-rails and toe-boards are not required.

(6)   Guard-rails and toe-boards are not required on: a ladder scaffold where a hand hold is provided; a trestle scaffold; a platform to be used only for a short time in assembling prefabricated structures, provided it is not less than 800 mm wide, has adequate hand hold, and is not used for materials unless they are kept in boxes to prevent them from falling; temporary platforms passing between glazing bars on a sloping roof where the provision of guard-rails and toe-boards is impractical, platforms under and suspended from a roof for light work of short duration but there must be adequate handholds and the material for the work must be

such that the platform can be used safely; and where working against a curved face provided steps are taken to prevent falls greater than 2 m.

(7) For work on sloping roofs (sloping more than 10 degrees) Regulation 35 applies.

Regulation 29: Guard-rails etc for gangway runs and stairs

(1), (2) and (3) Wherever a person could fall more than 2 m from a gangway, run or stairs, handrails 910 mm to 1.15 m must be provided on each side throughout their length and if necessary beyond. Except on stairs, toe-boards or barriers not less than 150 mm high must be fitted leaving no more than 765 mm between guard-rail and toe-board. Guard-rails and toe-boards may be removed or not put in place but only for so long as may be necessary for access and they need not be used on a temporary gangway used for work in assembling framework and only needed for a short time.

Regulation 30: Platforms, gangways, runs and stairs, etc to afford safe footholds

(1) and (2) All such must be kept free of obstruction and projecting nails and cleared or sanded as may be necessary should they become slippery.

Regulation 31: Construction and maintenance of ladders and folding step-ladders

(1), (2) and (3) Every ladder shall be properly constructed, suited for its use and properly maintained. Rungs must not be fixed to stiles by nails or the like alone.

Regulation 32: Use of ladders and folding step-ladders

(1) This Regulation applies to all ladders and folding ladders except for roof or duck ladders.

(2) Every ladder standing on its base must be: tied at the top or where that is impractical secured near the bottom; firmly seated equally on each stile never on loose bricks or packing; and supported intermediately where necessary to prevent sag or sway.

(3) Where a ladder cannot be tied it must be 'footed' by a person standing on the bottom rung for the whole time that any other person is using it.

(4) This Regulation does not apply to a ladder less than 3 m

long and not used as a means of communication provided it is securely placed.

(5) Ladders must extend 1.05 m above the landing place or top rung on which a person will stand unless there are other adequate handholds. There must be sufficient clear space behind each rung for adequate foothold.

(6) Suspended ladders must be securely and adequately suspended by each stile and supported intermediately where necessary to prevent sag or sway.

(7) Folding step-ladders must have firm and level footing never on loose bricks or packing.

(8) Wherever practicable intermediate landing places should be provided at intervals of not more than 30 feet where a ladder or ladders rise higher than this. The landing places must have guard-rails and toe-boards as for working platforms and any ladder opening in the floor should be as small as practicable.

Regulation 33: Openings, corners, breaks, edges and open joisting

(1) and (2)   Where a person may approach near or pass across any place where they would fall more than 2 m or into any liquid or material where there is risk of drowning or injury: then guard-rails and toe-boards must be provided as for working platforms; or the place must be covered or boarded over and the covering clearly marked or securely fixed.

(3)   Suitable precautions must be taken to prevent materials or articles falling so as to endanger persons.

(4)   Persons must be protected from falling more than 2 m through open joisting by securely fixed boarding or other effective means.

Regulation 34: Exceptions for Regulation 33

(1)   Guard-rails, toe-boards and coverings may be removed or not put in place but only for so long as may be necessary for access.

(2)   Demolition operations are exempted which are subject to Part X of the Construction (General Provisions) Regulations 1961 provided any place where persons could be endangered is not left unattended.

Regulation 35: Sloping roofs

(1)   Sloping roofs are any roof or part of a roof sloping more than

10 degrees from the horizontal, in the course of construction or in use for access to or egress from operations.

(2), (3) and (4) Where the roof or working conditions are dangerous or slippery or wherever the pitch is greater than 30 degress: only suitable workmen should do the work; sufficient, suitable and properly fixed crawl-boards should be provided; and either a barrier must prevent persons from falling over the edge or a working platform not less than 430 mm wide complete with guard-rails and toe-boards must be provided.

(5) Crawling ladders and boards must be properly constructed, maintained, supported and fixed.

(6), (7) and (8) Crawl-boards are not required where battens provide adequate foothold and handhold. The Regulations apply only where a person could fall more than 2 m but measures must be taken to ensure that materials and articles cannot fall.

Regulation 36: Work on or near fragile materials

(1) to (5) Where a person could fall more than 2 m through fragile material, duckboards or crawl-boards must be provided, securely fixed if necessary and sufficient to carry his weight plus any load he may carry. Where persons may work or pass near fragile materials guard-rail or suitable coverings must be provided. Prominent notices must warn those approaching of any fragile material except where it consists wholly of glass.

Regulation 37: Loads on scaffolds

(1) and (2) Scaffolds must be evenly loaded and not overloaded nor subjected to violent shock when loading. Only materials soon to be used should be put on a scaffold.

Regulation 38: Prevention of falls and provision of safety nets and belts

(1) to (4) Where it is not practicable to provide working platforms which comply with these Regulations then platforms shall be provided which do comply as far as is practicable and in addition properly maintained safety nets or sheets into which a person could fall without injury shall be provided. Where it is not practicable to provide such nets or sheets or where they have to be removed for the passage of materials or where the work is of short duration, then safety belts must be used. Safety belts must be

properly maintained, securely attached to suitable anchorage points or rails and together with their fittings designed to prevent injury in the event of a fall.

SCHEDULE   Regulation 22

FACTORIES ACT 1961

CONSTRUCTION (WORKING PLACES) REGULATIONS 1966

SCAFFOLD INSPECTIONS

FORM OF REPORTS OF RESULTS OF INSPECTIONS UNDER REGULATION 22 OF SCAFFOLDS, INCLUDING BOATSWAIN'S CHAIRS, CAGES, SKIPS AND SIMILAR PLANT OR EQUIPMENT (AND PLANT OR EQUIPMENT USED FOR THE PURPOSES THEREOF)

Name or title of Employer or Contractor...............................

Address of Site..

Work Commenced—Date..

Location and Description of Scaffold, etc. and other Plant or Equipment Inspected (1)	Date of Inspection (2)	Result of Inspection. State whether in good order (3)	Signature (or, in case where signature is not legally required, name) of person who made the inspection (4)

EXPLANATORY NOTE

(This Note is not part of the Regulations.)

For the protection of persons employed on building operations and on works of engineering construction, these Regulations impose requirements as to the safety of the working places and of the means of access and egress to and from those places. In the case of building operations these Regulations replace similar requirements in the Building (Safety, Health and Welfare) Regulations 1948.

Part IV: Keeping of records

Regulation 39: Reports, etc.

(1) and (2)   The reports required by Regulation 22 shall be kept available to any inspector or copies or extracts shall be sent to the inspector as he may require. They shall be kept on site except for jobs lasting six weeks or less when they may be kept in the contractor's office.

Schedule: Regulation 22

A specimen form for scaffold inspections is reproduced on page 145.

Appendix A

Definitions

Basic terms

Bay
> The space between the centre lines of two adjacent standards
> along the face of the scaffold.

Bay length
> The distance between the centres of two adjacent standards
> measured horizontally.

Height
> The height measured from the foundation to the top assembly of
> ledgers and transoms (cf. 'lift height').

Length
> The length of a scaffold between its extreme standards.
> Sometimes designated in number of bays (cf. 'bay length').

Lift
> The assembly of ledgers and transoms forming each horizontal
> level of a scaffold.

Foot lift
> A lift erected near to the ground. Sometimes known as a base lift.

Lift height
> The vertical distance between two lifts, measured centre to
> centre.

Lift head-room
> The clear distance between platform and the tubular assembly of
> the lift above.

Pair of standards
> The standards forming the frame at right angles to the building.

Scaffold
> A temporarily provided structure which provides access, or on or
> from which persons work, or which is used to support materials,
> plant or equipment.

Dead shore

    A vertical shore used to support a heavy point load, as in Falsework.

Cantilever drop scaffold

    An independent tied scaffold, erected on beams cantilevering out from a building. It is used in cases where it is impracticable or undesirable to found the scaffold on the ground.

Free-standing scaffold

    A scaffold which is not attached to any other structure and is stable against overturning on its own account or if necessary assisted by guys or rakers and anchors.

Independent tied scaffold

    A scaffold which has two lines of standards, one line supporting the outside of the deck and one the inside. The transoms are not built into the wall of the building. It is not free-standing, being supported by the building.

Putlog scaffold

    A scaffold which has one line of standards to support the outside edge of the deck and utilizes the wall built or the building to support the inside edge.

Raking shore

    An angled supporting scaffold, erected to provide support to a wall or building, often necessary whilst repairs or alterations are being carried out.

Slung scaffold

    A scaffold hanging on tubes, ropes or chains from a structure overhead. It is not capable of being moved or lowered.

Stack scaffold

    A scaffold erected usually on a pitched roof to provide access to a chimney stack.

Suspended scaffold

    A scaffold hanging on ropes which is capable of being suspended, or raised and lowered.

Width

    The width of a scaffold measured at right angles to the ledgers from centre to centre of the uprights or in the clear. Sometimes designated by the number of boards within the uprights and the number beyond the uprights on extended transoms.

Tubular members and beams

Brace

A tube placed diagonally with respect to the vertical or horizontal members of a scaffold and fixed to them to afford stability.

Cross brace

See 'ledger brace'.

Facade brace

A brace parallel to the face of the building. Sometimes known as a face brace.

Face brace

See 'facade brace'.

Knee brace

A brace across the corner of an opening in a scaffold to stiffen the angles or to stiffen the end support of a beam.

Ledger brace

A brace at right angles to the building. Sometimes known as a cross brace.

Longitudinal or sway brace

A brace generally in the plane of the longer dimension of the scaffold, particularly birdcages.

Plan brace

A brace in a horizontal plane.

Transverse brace

A brace generally in the plane of the shorter dimension of the scaffold.

Bridle

A horizontal tube fixed across an opening or parallel to the face of a building to support the inner end of a putlog transom or tie tube.

Butt tube

A short length of tube.

Butting tube

A tube which butts up against the facade of a building or other surface to prevent the scaffold moving towards that surface.

Chord

The principal longitudinal member(s) of a beam or truss.

Chord stiffener

A tube fixed at right angles to the chord of a prefabricated rafter, beam or truss for the purpose of preventing buckling.

Guardrail

A member incorporated in a structure to prevent the fall of a
person from a platform or access way.

End guardrail

A guardrail placed across the end of a scaffold or to isolate an
unboarded part.

Guardrail post

A puncheon supporting a guardrail.

Ledger

A longitudinal tube normally fixed parallel to the face of a
building in the direction of the larger dimensions of the scaffold.
It acts as a support for the putlogs and transoms and frequently
for the tie tubes and ledger braces, and is usually joined to the
adjacent standards.

Puncheon

A vertical tube supported at its lower end by another scaffold
tube or beam and not by the ground or on a deck.

Purlin

A tube secured to the rafters of a building parallel to the ridge for
the purpose of attaching the roof covering and to act as a top
chord stiffener for the rafter beams.

Putlog

A tube with a flattened end to rest in or on part of the brickwork
or structure.

Rafter and rafter beam

A transverse tube, beam or truss in a building spanning across a
roof or from the eaves to the ridge.

Raker

An inclined load-bearing tube.

Reveal tube

A tube fixed by means of a threaded fitting or by wedging
between two opposing surfaces of a structure, e.g. between two
window reveals, to form an anchor to which the scaffold may be
tied.

Sheeting rail

A horizontal tube fixed to the verticals of a scaffold to support the
sheeting.

Spine beam

A longitudinal main beam spanning from end to end of a roof at
the ridge or eaves.

Standard
> A vertical or near vertical tube.

Tie or tie assembly
> The components attached to an anchorage or the building or framed around a part of it or wedged or screwed into it with a tie tube. Used to secure the scaffold to the structure.

Bolted tie
> An assembly of nuts, bolts, anchors, rings or tubes fixed into the surface of a building.

Box tie
> An assembly of tubes and couplers forming a frame round a part of a building.

Lip tie
> An assembly of tubes forming an L or J-shaped hook round an inside surface of a building.

Double lip tie
> A lip tie which is a push and pull tie, i.e. has a cross tube on the back and front of the wall.

Movable ties
> Ties which may be temporarily moved for the execution of work. Two-way ties and reveal ties are two examples.

Non-movable ties
> Ties fixed to the structure where the face of the building is strong enough for their use, e.g. Hilti ties and Rawl-bolt ties.

Prop tie
> An assembly of telescopic props and/or scaffold tube jacked or wedged between the floors of a storey inside a building and including a tie tube.

Pull tie
> A tie which only acts to prevent the scaffold moving either towards or away from the building, e.g. a reveal tie, a box tie, a double lip tie, a bolted tie with a tie tube.

Reveal tie
> The assembly of a reveal tube with wedges or screwed fittings and pads if required, fixed between opposing faces of an opening in a wall together with the tie tube.

Through tie
> A tie assembly through a window or other opening in a wall.

Wire or band tie
> An assembly of a ring anchor and wire or steel banding used to tie the scaffold to the building.

Tie tube
> A tube used to connect a scaffold to an anchorage.

Transom
> A tube spanning across ledgers to form the support for boards or units forming the working platform or to connect the outer standards to the inner standards.

Butting transom
> A transom extended inwards to butt the building to prevent the scaffolding moving towards the building.

Needle transom
> A transom extending from or into a building.

Sway transom
> A transom extended inwards in contact with a reveal or the side of a column to prevent the scaffold moving sideways.

Scaffold couplers and fittings

Baseplate
> A metal plate with a spigot for distributing the load from a standard or raker or other load-bearing tube.

Adjustable baseplate
> A metal baseplate embodying a screw-jack.

Board clip
> A clip for fixing a board to a scaffold tube.

Toe-board clip
> A clip used for attaching toe-boards to tubes.

End toe-board clip
> A similar device for use on end toe-boards.

Coupler
> A component used to fix scaffold tubes together.

Check or safety coupler
> A coupler added to a joint under load to give security to the coupler(s) carrying the load.

Fixed finial coupler
> A fitting to fix a tube across the end of another at right angles in the same plane as in the hand-rail.

Load-bearing coupler
> A coupler having a rated load capacity.

Parallel coupler
> A coupler used to joint two tubes in parallel.

Purlin, rafter and ridge coupler
> Special angle or variable angle couplers for joining members in sheeted buildings and roofs.

Putlog coupler
> A coupler used for fixing a putlog or transom to a ledger.

Right-angle coupler
> A coupler used to join tubes at right angles.

Sleeve coupler
> An external coupler used to join one tube to another coaxially.

Supplementary coupler
> Coupler(s) added to a joint to back up the main coupler taking the load when the estimated load on the joint is in excess of the safe working load of the main coupler.

Swivel coupler
> A coupler used for joining tubes at an angle other than a right angle.

Swivel finial coupler
> A fitting to fix a tube across the end of another in the same plane but at an angle, like the hand-rail to a staircase.

Fittings
> A general term embracing components other than couplers.

Forkhead
> A U-shaped housing for assembly on the end of a tube to accept bearers.

Adjustable forkhead
> A forkhead fitted with a threaded spindle and nut to give adjustable height.

Rocking or swivel forkhead
> A forkhead to accept bearers at a range of angles.

Hop-up bracket or extension bracket
> A bracket to attach usually to the inside of a scaffold to enable boards to be placed between the scaffold and the building.

Joint pin
> See 'spigot'.

Putlog adaptor
> A fitting to provide a putlog blade on the end of scaffold tube.

Retaining bar
> A strip or device fixed across the top of the decking to hold it down.

Reveal pin
>A fitting used for tightening a reveal tube between two opposing surfaces.

Roofing clip or sheeting clip
>A fitting for fixing roof or wall sheeting to tubes in structures without the need for holes in the sheeting.

Sheeting hook
>A threaded rod hook with a washer and a nut used for attaching sheeting to tubes.

Soleplate
>A timber, concrete or metal spreader used to distribute the load from a standard or baseplate to the ground (also known as a sill).

Spigot
>An internal fitting to join one tube to another coaxially. Also known as a joint pin.

Expanding spigot
>A spigot incorporating an expanding device.

Fixed spigot
>A spigot permanently fixed to the end of a scaffold tube.

Spigot pin
>A pin placed transversely through the spigot and the scaffold tube to prevent the two from coming apart. Sometimes referred to as a dowel pin, tension pin, or interlock pin.

Sill
>See 'sole plate'.

Tension pin
>See 'spigot pin'.

Other terms in general use

Anchorage
>Component cast or fixed into the building for the purpose of attaching a tie.

Brick-guard
>A metal or other fender filling the gap between the guardrail and toe-board, and sometimes incorporating one or both of these components (sometimes referred to as 'retaining boards').

Castor
>A swivelling wheel secured to the base of a vertical member for the purpose of mobilizing the scaffold.

Gin wheel or block
> A single pulley for fibre ropes attached to a scaffold for raising or lowering materials.

Guy anchor
> A pin or tube driven into the ground at approximately 45° to the horizontal to provide an anchorage for a rope.

Inside board
> A board between the scaffold and the building on extended transoms, or a hop-up bracket.

Jib crane
> A small crane specially adapted for pivotal mounting to a scaffold tube.

Kentledge
> A dead-weight built in or added to a structure to ensure adequate stability.

Retaining boards
> See 'brick-guard'.

Scaffold board
> A softwood board generally used with similar boards to provide access, working platforms and protective components such as toe-board on a scaffold (or scaffold unit).

Scaffold or decking unit
> The board(s) or unit forming the working deck.

Sheeting
> Horizontal, vertical or inclined sheets of material, such as corrugated metal or plastic sheet, attached to a scaffold in order to provide protection from the effects of weather or alternatively to protect the surrounding area from the effects of works being carried out from the scaffold structure.

Skirt
> A short portion of vertical sheeting usually adjacent to the edge of a roof to give extra protection to the area enclosed immediately under the roof.

Toe-board
> An upstand at the edge of a platform intended to prevent materials or operatives feet from slipping off the platform.

End toe-board
> A toe-board at the end of a scaffold or at the end of a boarded portion of it.

Index